Internet-based Control Systems

Other titles published in this series:

Digital Controller Implementation and Fragility
Robert S.H. Istepanian and
James F. Whidborne (Eds.)

Optimisation of Industrial Processes at Supervisory Level
Doris Sáez, Aldo Cipriano and Andrzej W. Ordys

Robust Control of Diesel Ship Propulsion
Nikolaos Xiros

Hydraulic Servo-systems
Mohieddine Jelali and Andreas Kroll

Model-based Fault Diagnosis in Dynamic Systems Using Identification Techniques
Silvio Simani, Cesare Fantuzzi and Ron J. Patton

Strategies for Feedback Linearisation
Freddy Garces, Victor M. Becerra, Chandrasekhar Kambhampati and Kevin Warwick

Robust Autonomous Guidance
Alberto Isidori, Lorenzo Marconi and Andrea Serrani

Dynamic Modelling of Gas Turbines
Gennady G. Kulikov and Haydn A. Thompson (Eds.)

Control of Fuel Cell Power Systems
Jay T. Pukrushpan, Anna G. Stefanopoulou and Huei Peng

Fuzzy Logic, Identification and Predictive Control
Jairo Espinosa, Joos Vandewalle and Vincent Wertz

Optimal Real-time Control of Sewer Networks
Magdalene Marinaki and Markos Papageorgiou

Process Modelling for Control
Benoît Codrons

Computational Intelligence in Time Series Forecasting
Ajoy K. Palit and Dobrivoje Popovic

Modelling and Control of Mini-Flying Machines
Pedro Castillo, Rogelio Lozano and Alejandro Dzul

Ship Motion Control
Tristan Perez

Hard Disk Drive Servo Systems (2nd Ed.)
Ben M. Chen, Tong H. Lee, Kemao Peng and Venkatakrishnan Venkataramanan

Measurement, Control, and Communication Using IEEE 1588
John C. Eidson

Piezoelectric Transducers for Vibration Control and Damping
S.O. Reza Moheimani and Andrew J. Fleming

Manufacturing Systems Control Design
Stjepan Bogdan, Frank L. Lewis, Zdenko Kovačić and José Mireles Jr.

Windup in Control
Peter Hippe

Nonlinear H_2/H_∞ Constrained Feedback Control
Murad Abu-Khalaf, Jie Huang and Frank L. Lewis

Practical Grey-box Process Identification
Torsten Bohlin

Control of Traffic Systems in Buildings
Sandor Markon, Hajime Kita, Hiroshi Kise and Thomas Bartz-Beielstein

Wind Turbine Control Systems
Fernando D. Bianchi, Hernán De Battista and Ricardo J. Mantz

Advanced Fuzzy Logic Technologies in Industrial Applications
Ying Bai, Hanqi Zhuang and Dali Wang (Eds.)

Practical PID Control
Antonio Visioli

(continued after Index)

Shuang-Hua Yang

Internet-based Control Systems

Design and Applications

Shuang-Hua Yang
Department of Computer Science
Loughborough University
Holywell Park
Loughborough LE11 3TU
United Kingdom
s.h.yang@lboro.ac.uk

ISSN 1430-9491
ISBN 978-1-84996-358-9   e-ISBN 978-1-84996-359-6
DOI 10.1007/978-1-84996-359-6
Springer London Dordrecht Heidelberg New York

Library of Congress Control Number: 2011920993

© Springer-Verlag London Limited 2011

Sun, Sun Microsystems, the Sun Logo and Java are trademarks or registered trademarks of Sun Microsystems, Inc. in the United States and other countries.

MATLAB®, Simulink® and Symbolic Math Toolbox™ are trademarks of The MathWorks, Inc., 3 Apple Hill Drive, Natick, MA, 01760-2098 USA, http://www.mathworks.com

LabVIEW™ is a trademark of National Instruments. National Instruments Corporation, 11500 N Mopac Expwy, Austin, TX 78759-3504, U.S.A. http://www.ni.com

WinCC is a registered trademark of Siemens Aktiengesellschaft, Wittelsbacherplatz 2, 80333 Munich, Germany HYPERLINK http://www.siemens.com

Apart from any fair dealing for the purposes of research or private study, or criticism or review, as permitted under the Copyright, Designs and Patents Act 1988, this publication may only be reproduced, stored or transmitted, in any form or by any means, with the prior permission in writing of the publishers, or in the case of reprographic reproduction in accordance with the terms of licenses issued by the Copyright Licensing Agency. Enquiries concerning reproduction outside those terms should be sent to the publishers.

The use of registered names, trademarks, etc., in this publication does not imply, even in the absence of a specific statement, that such names are exempt from the relevant laws and regulations and therefore free for general use.

The publisher makes no representation, express or implied, with regard to the accuracy of the information contained in this book and cannot accept any legal responsibility or liability for any errors or omissions that may be made.

Printed on acid-free paper

Springer is part of Springer Science+Business Media (www.springer.com)

For Lili, my daring wife, Bob and James, my two handsome sons. I love you more than you can imagine.

Advances in Industrial Control

Series Editors

Professor Michael J. Grimble, Professor of Industrial Systems and Director
Professor Michael A. Johnson, Professor (Emeritus) of Control Systems
and Deputy Director

Industrial Control Centre
Department of Electronic and Electrical Engineering
University of Strathclyde
Graham Hills Building
50 George Street
Glasgow G1 1QE
United Kingdom

Series Advisory Board

Professor E.F. Camacho
Escuela Superior de Ingenieros
Universidad de Sevilla
Camino de los Descubrimientos s/n
41092 Sevilla
Spain

Professor S. Engell
Lehrstuhl für Anlagensteuerungstechnik
Fachbereich Chemietechnik
Universität Dortmund
44221 Dortmund
Germany

Professor G. Goodwin
Department of Electrical and Computer Engineering
The University of Newcastle
Callaghan
NSW 2308
Australia

Professor T.J. Harris
Department of Chemical Engineering
Queen's University
Kingston, Ontario
K7L 3N6
Canada

Professor T.H. Lee
Department of Electrical and Computer Engineering
National University of Singapore
4 Engineering Drive 3, Singapore 117576

Professor (Emeritus) O.P. Malik
Department of Electrical and Computer Engineering
University of Calgary
2500, University Drive, NW
Calgary, Alberta
T2N 1N4, Canada

Professor K.-F. Man
Electronic Engineering Department
City University of Hong Kong
Tat Chee Avenue
Kowloon, Hong Kong

Professor G. Olsson
Department of Industrial Electrical Engineering and Automation
Lund Institute of Technology
Box 118, S-221 00 Lund
Sweden

Professor A. Ray
Department of Mechanical Engineering
Pennsylvania State University
0329 Reber Building
University Park
PA 16802, USA

Professor D.E. Seborg
Chemical Engineering
3335 Engineering II
University of California Santa Barbara
Santa Barbara
CA 93106, USA

Doctor K.K. Tan
Department of Electrical and Computer Engineering
National University of Singapore
4 Engineering Drive 3, Singapore 117576

Professor I. Yamamoto
Department of Mechanical Systems and Environmental Engineering
The University of Kitakyushu
Faculty of Environmental Engineering
1-1, Hibikino,Wakamatsu-ku, Kitakyushu, Fukuoka, 808-0135
Japan

Series Editors' Foreword

The series *Advances in Industrial Control* aims to report and encourage technology transfer in control engineering. The rapid development of control technology has an impact on all areas of the control discipline. New theory, new controllers, actuators, sensors, new industrial processes, computer methods, new applications, new philosophies,..., new challenges. Much of this development work resides in industrial reports, feasibility study papers and the reports of advanced collaborative projects. The series offers an opportunity for researchers to present an extended exposition of such new work in all aspects of industrial control for wider and rapid dissemination.

The Internet and the World Wide Web (WWW) have created a revolution in the way information is archived, accessed and processed. Furthermore, these systems have wrought changes in communication modalities that have been no less profound. Business, commercial and societal interactions have all changed irrevocably. The ways in which the control systems community works have undergone comparable changes, ranging from the new mechanisms societies like the IEEE Control System Society use to communicate with their members, run their conferences and publish their conference proceedings and journals through to the actualities of remotely controlling industrial and laboratory processes. It is the impact and potential that the Internet has for control system design, implementation and operation that is the subject of this new *Advances in Industrial Control* monograph entitled *Internet-based Control Systems: Design and Applications* by Professor Shuang-Hua Yang of Loughborough University in the UK.

The monograph is very interesting to read and is comprehensive in its coverage of "control using the Internet". It reports on the characteristics of Internet data and on the features of Internet communications and discusses control architectures, control design procedures and implementation concepts for use with Internet-based control systems. The monograph also contains applied case studies of the concepts presented and ideas proposed by the author. Two issues emerge from the text:

- The Internet as a process
- Design and implementation for control systems that use the Internet

Among the characteristics and special features of the Internet as a process, we read about:

- Time delays that are unknown and mostly associated with data transfer times
- Information disorder and loss, so that packets of data can arrive at a destination out of time order, or may not even arrive at all
- Security, where hacking into data processes to change or disrupt control actions is always a possibility
- Safety, involving the issues of remotely monitoring and promoting the safe and fault-free operations of control systems and their controlled processes

Professor Yang deals with all these aspects in the monograph, prescribing ways to improve and enhance the reliability of the Internet process. Moving on to the second issue of design and implementation for control systems that incorporate a unit or block called "the Internet", Professor Yang investigates key areas including random time delays, mixed continuous-discrete systems, hierarchical control architectures and multi-rate sampled systems. Where implementation is concerned, safety and safety issues play an important role and ideas from the domain of performance monitoring and controller assessment are used. In the monograph, both simulation examples and real application case studies are reported to provide convincing evidence of the viability, soundness and performance of the control designs presented and the implementation strategies proposed by the author.

The topicality, immediacy and comprehensiveness of this monograph are likely to be appreciated by a wide range of readers. Industrial engineers from a broad set of disciplines will find the monograph can help to provide answers to questions like, "How can using the Internet help a company control, and optimise its processes for better economic and technical performance?" or "What measures does a company need to take to protect Internet-based control schemes?". Many academics and students may already have experienced using the Internet to control and experiment with laboratory equipment based at geographically distant locations. This monograph can provide an overview of and detailed insights into the techniques used in such endeavours, and an entry for those unfamiliar with the techniques involved. Consequently, a control community and educational readership should find the monograph of interest, too.

The Editors are very pleased to have this monograph enter the series as it well demonstrates and offers the real prospects of advances in industrial control technology.

Industrial Control Centre M.J. Grimble
Glasgow M.A. Johnson
Scotland, UK
2010

Preface

Nowadays the Internet plays a very important role not only in our daily life and work, but also in real-time industrial manufacturing, scheduling and management. During the last decade, considerable research has been carried out to develop new technologies, called Internet-based control systems in our research, that make it possible to supervise and control industrial processes over the Internet. The use of the Internet for real-time interaction and for the remote control and monitoring of plant would bring many benefits to industry. Although such an approach seems promising, there remain many new challenges raised by the introduction of the Internet into control systems. The design methods of traditional control systems appear not to be completely suitable for this new type of control system. Consequently, new design issues need to be considered including requirement specification, architecture selection, user interface design, Internet latency, networked control system stability, safety and security. For example, Internet transmission delay can lead to irregular data transmission and data loss. In the worst case, this can make the whole system unstable. We also need to address the problem of security. If malicious hackers gain access to an Internet-enabled control system, the consequences could be catastrophic.

This book is concerned with the study of Internet-based control systems, which are systems by which sensors and actuators in different locations may be controlled and monitored from a central hub, using the Internet as the communications network. The book aims to systematically present the methods developed by the author for dealing with all the above design issues for Internet-based control systems. Furthermore, the application perspectives of Internet-based control are explored through a number of application systems. The book contains the latest research, much of which the author has performed or been intimately involved with, which presents new solutions to existing problems and explores the manifold future applications of Internet-based control systems. This book is unique in bringing together multiple strands of research, mainly from computer science and control engineering, into an over-arching study of the entire subject. Every time when I talk to industrial people about Internet-based control systems, the immediate question is

"Is it really worth doing in view of the safety risk of Internet-enabled control systems". This book may build your confidence in fully exploiting the new technologies described here in your research and/or industrial work.

This book consists of 13 chapters, including an introduction and conclusion. Chapters 2–10 focus on providing solutions to the above design issues from a control engineering, or computer network perspective or from a fusion of the two. Chapters 11 and 12 explore the future applications of Internet-based control systems in remote performance monitoring, remote design, testing and maintenance.

The book can serve both as a textbook and a reference book. The potential audience for this book includes researchers in control engineering and computer networks, control and system engineers, real-time control system software developers and IT professionals. This book can also be used as a textbook for a final year option or elective on Internet-enabled system design, or as an advanced example of real-time software design at the postgraduate level. It can also be used as a textbook for teaching Internet-enabled systems in general.

Loughborough, UK   Shuang-Hua Yang

Acknowledgements

Many people have directly or indirectly contributed to the work presented in this book. I would like to thank my former research assistant, Dr Xi Chen, my former PhD students, Dr Chengwei Dai and Dr Yunquan Li, for working with me in this area. I would also like to thank Prof. Qing-Guo Wang at National University of Singapore, Prof. Yongji Wang and Prof. Zhi-Hong Guan at Huazhong University of Science and Technology, China, Prof. Liansheng Tan at Central China Normal University, China, Prof. Xuemin Tian, Prof. Bokia Xia and Dr Yu-Hong Wang at Petroleum University, China, Prof. Guoping Liu at Glamorgan University, Dr Yi Cao at Cranfield University, Prof. Paul Chung, Prof. James Alty, Dr Lili Yang, Dr Wen-Hua Chen and Dr Walter Hussak at Loughborough University, Dr Alan Grigg, Dr Julian Johnson, Dr Brian Ford and Dr Tony Martin at the BAE Systems, with whom I collaborated over the years. My appreciation also goes to my colleagues at the Computer Science Department at Loughborough University for their enthusiasm and dedicated assistance they have provided for me.

My gratitude goes to Prof. Michael Johnson, the series editor, for his encouragement and comments to the manuscript, to my colleague, Dr Roger Knott, and Ms Charlotte Cross (Springer-Verlag) for their proof reading, to my exchange PhD student, Mr Wei Zheng, for his graphic expertise and to Ms Yanning Yang, my PhD student, Ms Lu Tian, a Loughborough graduate, and Mr Thomas Welch, my former personal assistant, for their editorial work.

Finally, I gratefully acknowledge the financial supports from the Engineering and Physical Sciences Research Council (EPSRC) in the UK, the Royal Society Research Grant, the Royal Society incoming visiting professorship, the Royal Academy of Engineering Industrial Secondment Award, Loughborough University Research Studentship and the Hong Kong Wang KuanCheng Foundation. Even though I do not refer to the related work we did for the Ministry of Defence (MoD) in the book for confidential reasons, the research grant from the MoD through the Systems Software Engineering Initiative (SSEI) programme has enabled us to significantly extending the use of the Internet-based control systems from process industry to military applications and is deeply appreciated.

March 2010

Contents

1 Introduction........ 1
   1.1 Networked Control Systems (NCS)........ 1
   1.2 Internet-based Control Systems (ICS)........ 2
   1.3 Challenges of NCS/ICS........ 3
   1.4 Aims of the Book........ 4
   References........ 5

2 Requirements Specification for Internet-based Control Systems........ 7
   2.1 Introduction........ 7
   2.2 Requirements Specification........ 7
   2.3 Functional Modelling of Internet-based Control Systems........ 9
   2.4 Information Hierarchy........ 12
   2.5 Possible Implementation of Information Architecture........ 14
   2.6 Summary........ 15
   References........ 16

3 Internet-based Control System Architecture Design........ 17
   3.1 Introduction........ 17
   3.2 Traditional Bilateral Tele-operation Systems........ 17
   3.3 Remote Control over the Internet........ 21
   3.4 Canonical Internet-based Control System Structures........ 24
   3.5 Summary........ 26
   References........ 26

4 Web-based User Interface Design........ 29
   4.1 Features of Web-based User Interface........ 29
   4.2 Multimedia User Interface Design........ 29
   4.3 Case Study........ 31
      4.3.1 System Architecture........ 31
      4.3.2 Design Principles........ 33

		4.3.3	Implementation	34
	4.4	Summary		35
	References			35

5 Real-time Data Transfer over the Internet ... 37
- 5.1 Real-time Data Processing ... 37
  - 5.1.1 Features of Real-time Data Transfer ... 38
  - 5.1.2 Light and Heavy Data ... 38
- 5.2 Data Wrapped with XML ... 40
  - 5.2.1 Structure Mapping ... 40
  - 5.2.2 Data Mapping ... 42
- 5.3 Real-time Data Transfer Mechanism ... 42
  - 5.3.1 RMI-based Data Transfer Structure ... 42
  - 5.3.2 Data Object Priority ... 44
- 5.4 Case Study ... 45
  - 5.4.1 System Description ... 45
  - 5.4.2 Priority of Data Transfer ... 47
  - 5.4.3 Implementation ... 47
  - 5.4.4 Simulation Results and Analysis ... 48
  - 5.4.5 Advantages of RMI-based Data Transfer ... 50
- 5.5 Summary ... 51
- References ... 52

6 Dealing with Internet Transmission Delay and Data Loss from the Network View ... 53
- 6.1 Requirements of Network Infrastructure for Internet-based Control ... 53
  - 6.1.1 Six Requirements for Ideal Network Infrastructure for Internet-based Control ... 53
- 6.2 Features of Internet Communication ... 54
- 6.3 Comparison of TCP and UDP ... 55
- 6.4 Network Infrastructure for Internet-based Control ... 56
  - 6.4.1 Real-time Control Protocol ... 57
  - 6.4.2 Quality Service Provider and Time Synchronization ... 59
- 6.5 Typical Implementation for Internet-based Control ... 60
  - 6.5.1 Experimental Set-up ... 60
  - 6.5.2 Implementation ... 62
- 6.6 Summary ... 64
- References ... 65

7 Dealing with Internet Transmission Delay and Data Loss from the Control Perspective ... 67
- 7.1 Overcoming the Internet Transmission Delay ... 67
- 7.2 Control Structure with the Operator Located Remotely ... 68

	7.3	Internet-based Control with a Variable Sampling Time	69
	7.4	Multi-rate Control	71
		7.4.1 Two-level Hierarchy in Process Control	71
		7.4.2 Multi-rate Control	72
	7.5	Time Delay Compensator Design	73
		7.5.1 Compensation at the Feedback Channel	76
		7.5.2 Compensation at the Feed-forward Channel	77
	7.6	Simulation Studies	78
		7.6.1 Simulation of Multi-rate Control Scheme	78
		7.6.2 Simulation of Time Delay Compensation with a Variable Sampling Time	79
	7.7	Experimental Studies	85
		7.7.1 Virtual Supervision Parameter Control	85
		7.7.2 Dual-rate Control with Time Delay Compensation	91
	7.8	Summary	95
	References	96
8	**Design of Multi-rate SISO Internet-based Control Systems**	99
	8.1	Introduction	99
	8.2	Discrete-time Multi-rate Control Scheme	100
	8.3	Design Method	101
	8.4	Stability Analysis	104
	8.5	Simulation Studies	105
	8.6	Real-time Implementation	107
	8.7	Summary	110
	References	111
9	**Design of Multi-rate MIMO Internet-based Control Systems**	113
	9.1	Introduction	113
	9.2	System Modeling	114
		9.2.1 State Feedback Control	114
		9.2.2 Output Feedback Control	115
	9.3	Controller Design	116
	9.4	Stability Analysis	118
	9.5	Design Procedure	121
	9.6	Model-based Time Delay Compensation	121
		9.6.1 Compensation of the Transmission Delay at the Feedback Channel	123
		9.6.2 Compensation of the Transmission Delay in the Feed-forward Channel	124
		9.6.3 Unified Form of the State Feedback Control of the Remote Controller	125
	9.7	Simulation Study	125
	9.8	Summary	127
	References	128

10 Safety and Security Checking ... 131
- 10.1 Introduction ... 131
- 10.2 Similarity of Safety and Security ... 132
- 10.3 Framework of Security Checking ... 132
  - 10.3.1 Framework of Stopping Possible Malicious Attack ... 132
  - 10.3.2 Framework-based What-If Security Checking ... 134
- 10.4 Control Command Transmission Security ... 136
  - 10.4.1 Hybrid Algorithm ... 136
  - 10.4.2 Experimental Study ... 138
- 10.5 Safety Checking ... 139
- 10.6 Case Study ... 141
  - 10.6.1 Ensuring Security ... 142
  - 10.6.2 Safety Checking ... 142
- 10.7 Summary ... 144
- References ... 144

11 Remote Control Performance Monitoring and Maintenance over the Internet ... 147
- 11.1 Introduction ... 147
- 11.2 Performance Monitoring ... 148
  - 11.2.1 Acquisition and Storage of Data ... 149
  - 11.2.2 Data Analysis and Performance Identification ... 149
  - 11.2.3 Categories of Performance Monitoring ... 150
- 11.3 Performance Monitoring of Control Systems ... 151
  - 11.3.1 General Guidelines of Control Performance Monitoring ... 151
  - 11.3.2 Control Performance Index and General Likelihood Test ... 152
  - 11.3.3 Performance Compensator Design ... 155
- 11.4 Remote Control Performance Maintenance ... 156
  - 11.4.1 Architecture of Remote Maintenance ... 156
  - 11.4.2 Implementation of Back-end System ... 158
  - 11.4.3 Implementation of Front-end System ... 159
- 11.5 Case Study ... 161
  - 11.5.1 System Description ... 161
  - 11.5.2 Setting up a Fault ... 162
  - 11.5.3 Fault Compensation ... 164
- 11.6 Summary ... 165
- References ... 166

12 Remote Control System Design and Implementation over the Internet ... 169
- 12.1 Introduction ... 169
- 12.2 Real-time Control System Life Cycle ... 170

	12.3	Integrated Environments	171
		12.3.1 Interaction Between Real World and Virtual World	171
		12.3.2 Available Integrated Frameworks	173
		12.3.3 Architecture of a General Integrated Environment	177
	12.4	A Typical Implementation of the General Integrated Environment	178
		12.4.1 Design Workbench	180
		12.4.2 Implementing a New Design of a Controller	182
		12.4.3 Collaboration in the Integrated Environment	185
	12.5	Case Study	187
		12.5.1 Workbench for Testing	188
		12.5.2 Testing the Model and the Controller of the Water Tank at the Workbench	188
		12.5.3 Installation of the New Design of Real Controllers	190
	12.6	Summary	192
	References		193
13	**Conclusion**		195
	13.1	Summary	195
	13.2	Future Work	196
	References		197
Index			199

Chapter 1
Introduction

1.1 Networked Control Systems (NCS)

The Internet has shown itself to be the most successful communications network of the past decade, allowing a level of information sharing previously unimaginable. Online libraries can be accessed in seconds and video and audio files can be shared between millions of consumers in an instant. Its popularity as a tool for information dissemination and communication has led to it being ubiquitous across the globe, with more and more devices possessing Internet connectivity. This makes it an incredibly flexible network as there is almost always an access point close at hand and, as it is an existing network, it is less expensive to use than other networks. These factors have made the Internet the ideal communications system for both government and industry to use in controlling objects as diverse as robotic laboratories and industrial plants in different locations around the world. Much like the networked control systems (NCS) that came before, Internet-based control promises a great many benefits for government, industry, and individuals alike.

NCS as a research area is incredibly important and promises many new advances. It is the study of how to use existing communication networks to link controllers or control centres with actuators and sensors that may be in geographically disparate locations. Such networks may be hard wired, such as through a phone line, or wireless, such as the Bluetooth and Wi-Fi systems. It is important to distinguish between an NCS, which is control *over* a network, and control *of* a network, which aims at improving quality of services.

In the early development of control systems, the principal structure was a centralized control structure; the controller, the sensors and actuators of which had to be point-to-point wired and physically located in a short distance. While this resulted in no time delay in signal transfer and no signal loss, it was very expensive. In the 1970s and 1980s, there was an increasing use of control systems in distributed large-scale systems, which were too large and complicated for the implementation of a centralized control structure. A manufacturing plant would be split into subsystems, each with their own controller and with no signal transfer between these separate subsystems. However, this decentralized control system was markedly inefficient and could not produce the required performance, leading to the creation of quasi-decentralized control systems (Gajic 1987) in and around the 1990s. These were, as the name suggests, a mixture of decentralized and centralized control systems; most signals were collected and processed locally; however, some were shared between local plants and remote controllers. At the time, remote-communication for online control was possible but exorbitant in cost, which kept the shared signals to a minimum.

NCS provided a means of returning to the efficient but expensive centralized control systems by using existing, shared communication networks to massively reduce cost and to allow access from other points to which the network is connected. Whilst using shared communications networks carries the risk of over-congestion leading to signal delay and data packet dropout, this risk is not inherent and, with improvements in communications networks such as Quality of Service (QoS), can be significantly reduced.

1.2 Internet-based Control Systems (ICS)

Internet-based Control Systems (ICS) (Yang 2005) are NCS that use the Internet as the shared communications network. It is obvious that such control systems could be incredibly useful as so many devices are Internet enabled – from phones to watches and PDAs to MP3 players. Not only the ability for these items to connect to the Internet has grown but the number of people possessing them has also expanded, which means more people have more items that connect them to NCS. Internet technologies enable the swift dissemination of information – such as variations in global markets, but, with ICS, it is theoretically possible for someone to directly control, monitor, and maintain a manufacturing plant in Beijing with a PDA in Swansea. Whilst this may seem somewhat absurd and even slightly worrying, it gives the ability for transnational companies to site the majority of their control engineers in one central facility in, say, New York, but be in control of systems in China, India, and other places where manufacturing costs are lower, which leads to massive saving on flying experts to diverse locations and allowing direct adjustments to changing market demands.

As with all remote control networks, ICS have certain key components:

- A control point from which commands may be issued and data may be monitored.
- A point at which those commands are carried out and data collected, often involving sensors and actuators.
- The Internet that communicates between the two.

Whilst no remote control network can work without these components, there are many that are far more complicated, with multiple control points, sensors and actuators, and some using different types of communication networks. As has been previously mentioned, the success of the Internet has been astounding. It has made a significant impact on society through its use as a communication and data transfer mechanism. Many systems are being created all over the world to implement Internet applications. Most of them are focused on tele-robotic systems. The creation of virtual laboratories over the Internet for educational purposes is also one of the areas that is being currently developed. Normally, we call the NCS involved with the Internet "ICS". Research on ICS is focused on guiding the design process, dealing with Internet latency, and ensuring safety and security. ICS creates a new window of opportunity for control engineers and has a range of practical applications, from monitoring military personnel and equipment to allowing consumers to see manufacturing conditions inside a plant in real time.

In terms of industrial manufacturing, it can provide companies with much greater flexibility as control engineers and specialists can all be sited in one place and command and monitor plants in diverse locations, saving on the time and expenditure entailed in flying out teams of experts. As the Internet has a vast number of access points, it allows skilled and experienced plant managers in one plant to give remote assistance to other plant managers and collaborate together even though they may be in different locations and none of them may be at the actual plant itself. It also gives the ability for sales and marketing teams to change production rates based on demand and market fluctuations. It may also be used to increase transparency – government watchdog agencies and consumer groups could monitor conditions within a plant to ensure best practice through public scrutiny.

Beyond the world of manufacturing, ICS may be used in collaborative research and educational projects, allowing multiple groups in universities, schools, and research facilities to share laboratories and carry out experiments for which they do not have the equipment to do themselves.

1.3 Challenges of NCS/ICS

There are a number of challenges that NCS and ICS (Yang and Cao 2008) must face regardless of whether NCS and ICS are established on wired or wireless networks. Three major challenges are network latency, safety and security, and multiple user access. The data transmission latency is the main difference between NCS/ICS and other tele-operations. Most tele-operating systems are based on private media, by which the transmission delay can be well modelled. The Internet, in contrast, is a public and shared resource in which various end users transmit data via the network simultaneously. The route for transmission between two end points in a wide area is not fixed for different paths and a traffic jam may be caused when too many users traverse the same route simultaneously. The transmission latency of any public network is difficult to model and predict.

Although introducing a public communication network such as the Internet into control systems has yielded benefits through the open architecture, there are various risks, which result from the application of network technologies. One typical risk is operational difficulties caused by cyber terrorists, especially in systems linked to the military or potentially dangerous industrial operations. The Internet is always open to hackers because of its open architecture. The hackers will try to cause failures by triggering them through the Internet using the open architectures. Therefore the scope of the ICS-controlled plant safety cannot be limited to within plant sites, because there is a possibility that the local control system would be accessed or falsified by outsiders through the Internet. On the other hand, authorized remote users may cause failures in the plant as well due to the Internet environment constraints.

Compared with local control systems, the special features of NCS and ICS are the multiple users and the uncertainty about who the users are, how many users there are, and where they are. In NCS and ICS, the operators cannot see each other or may never have met. It is likely that multi-users may try to concurrently control a particular parameter. If authorized users have the same opportunity to fully control the whole plant some problems could arise. Some mechanism is required to solve control conflict problems between multiple users and coordinate their operations.

Although the notion of NCS and ICS is still in its infancy in the academic control community, it is not new to industry. It has been used in manufacturing plants, aircraft, automobiles, health care, disaster rescue, etc. A number of automation and instrument companies have made their hardware and/or software products Internet enabled. For example, the latest version of the LabView™ produced by the National Instrument (NI) has a function for remote control over the Internet. Intuitive Technology Corp. provided "web @ aGlance" for feeding real-time data to an Internet-enabled Java graphics console. WinCC®, from SIEMENS, is able to link local control systems with the Internet. Sun Microsystems, Cyberonix, Foxboro, Valmet, Emerson, and Honeywell all have also made their products Internet enabled.

However, if you ask industry how to choose a NCS and ICS architecture, how to overcome the data transmission time delay and data loss, and how to ensure the safety and security for the Internet-enabled control systems, the answer you receive might not be very convincing. The reason is simple. These fundamental issues in the design of NCS and ICS are currently under investigation in the academic control community and there are no available convincing answers to them, although they have captured the interest of many researchers worldwide.

ICS suffers from the fact that it is communicating through a network, which was not designed to carry so much traffic – telephone landlines. This leads to multiple problems in terms of congestion appearing as network delay, sampling times, jitter, data packet dropout, and network scheduling. These are usually considered the preserve of Control of Communications Networks and they have a very obvious impact on the performance of ICS. This book shows how to deal with these problems from both control engineering and computer science aspects.

1.4 Aims of the Book

This book is intended as a reference book or textbook for undergraduate and postgraduate students as well as researchers of control engineering and information technologies. It is also useful for control and system engineers, company managers, real-time Internet enabling software developers, and IT professionals. Thus, it sets out to explore and examine the theory and applications of ICS, looking at specific examples and current research in the area and also at the future of ICS. This book confines itself to examining the purely technical side of the ICS and does not explore the ethical implications, such as the increased power it may give to governments, and corporations through rigid centralization, or the opportunity for terrorists to cause serious damage to delicate and toxic manufacturing processes. However, these may be important considerations for anyone intending to implement ICS.

References

Gajic, Z., (1987) On the quasi-decentralized estimation and control of linear stochastic systems, Systems and Control Letter, 8, pp. 441–444.

Yang, S.H., (2005) Internet-based control: the next generation of control systems, Measurement and Control, 38(1), pp. 11.

Yang, S.H., and Cao, Y., (2008) Networked control systems and wireless sensor networks: theories and applications, International Journal of Systems Science, 39(11), pp. 1041–1044.

Chapter 2
Requirements Specification for Internet-based Control Systems

2.1 Introduction

The design process for the development of Internet-based control systems includes requirement specification, architecture design, control algorithm design, interface design, and security and safety analysis. Specifying requirements for Internet-based control systems is the first task in the design process because different requirements may lead to different control architectures. The requirements specification should be met by the architecture design. Many requirements validation techniques exist, ranging from the building of prototypes or executable specifications to waiting until the system is constructed, and then testing the whole system. This latter approach leaves any testing until it is too late and too expensive to make any change in the specification of the control system, although certainly much can be learned by "testing" the specification (Levson et al. 1994). In this chapter, we discuss a way to systematically specify requirements for Internet-based control systems and to build a functional model to describe the requirements specification and then extend this functional model into an information hierarchy. The information hierarchy gives an indication to the architecture of the Internet-based control systems.

2.2 Requirements Specification

Generally, a system comprises a set of components working together to achieve some common purpose or objective. For a control system, the goal is to maintain a particular relationship between the inputs to the system and the outputs from the system in the face of disturbances in the process.

In addition to the basic objective required by the process, such systems may also have constraints on their operating conditions. Constraints limit the set of acceptable designs by which the objectives may be achieved and may arise from several sources including physical limitations and safety considerations.

As a result of such system constraints, the goals of a system may not be entirely achievable. These unachievable goals cannot be the requirements since it is not possible to achieve them. The major task in the requirements specification is to identify and resolve trade-offs between functional goals and constraints, which may cause goals to be not completely achievable. For Internet-based control systems, the requirements should include process monitoring and control objectives. Some requirements, however, require a deterministic timing regime and therefore may not be achievable due to Web-related traffic delay. Such requirements, which are not entirely achievable over the Internet, should be excluded.

We use process control as an example here to describe the procedure for specifying requirements for an Internet-based process control system. As shown in Fig. 2.1, process control system hierarchy is, in general, composed of management, plant-wide optimization, supervisory, regulatory, and sensor/actuator levels. The hypothesis advocated is that the possible requirements for such Internet-based process control are solely composed of those requirements that are achievable through the additional Internet control in the process control hierarchy. The process of the requirements specification is to refine and formalize the preliminary description of the goals, into a formal description, that is a requirement. The process can be classified into horizontal and the vertical dimensions. In the horizontal dimension, the goals split into sub-goals at the different levels. Through formalization, these

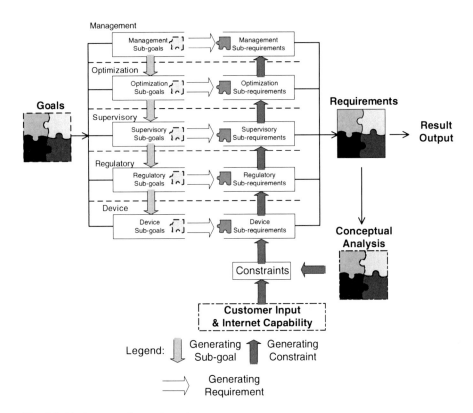

Fig. 2.1 Goals, constraints and requirements

sub-goals generate the sub-requirements corresponding to the control levels. The system requirements are obtained by integrating these sub-requirements together. In the vertical dimension, since the higher-level sub-goals need some support from low-level ones, they may generate the sub-goals and add these sub-goals to the low level, which can be considered as a top-down process. On the other hand, through conceptual analysis of the initial requirements and consideration of the Internet capability and customer inputs, constraints are generated to refine the initial requirements. The generated constraints propagate from bottom to top to refine the requirements. Through these processes, the final requirements can guarantee the maximum likelihood of achieving the goals subject to the constraints.

2.3 Functional Modelling of Internet-based Control Systems

In order to represent requirements specification, a functional model is required. Functional modelling consists of making a specification of a control system as expected by the user, which expresses what the control system has to do to allow the process to be controlled. These user needs are usually represented by Data Flow Diagrams (DFD) identifying control functions on the basis of the {B, I, C} triplet (Iung et al. 2001) as shown in Fig. 2.2, where

- **B** defines the behaviour of the function. It is modelled by function algorithms.
- **I** defines information produced or consumed by the control system of the function. It is modelled by the function interfaces.
- **C** defines communication. It is modelled by means of data flows connecting the functions with other functions within the system.

The structuring of the DFDs should allow designers to identify an actuation channel, to which an operator can send requests to fulfil the system objective, and a measurement channel, which reports the effects of these requests on the system.

The objective of establishing Internet-based control systems is to enhance rather than replace ordinary computer-based control systems by adding an extra Internet level to the control system (Yang et al. 2002, 2003). A pictorial schematic of an Internet-based control system is shown in Fig. 2.3. The Internet-based control system provides a means of monitoring and adjusting process plants from any

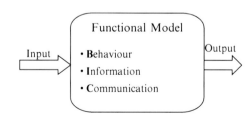

Fig. 2.2 System functional modelling

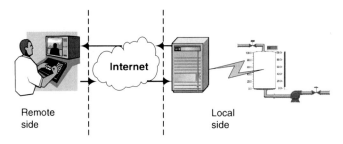

Fig. 2.3 Pictorial schematic of Internet-based control systems

location via the Internet using standard Web browsers and PCs. Obvious advantages of this approach include the following:

- Access to the monitoring functionality regardless of location
- Use of zero cost software (standard Web browsers) on the client side to access the information
- Internally or externally collaboration between participants

In order to describe the functions of Internet-based control systems in terms of the {B, I, C} triplet, a Process Control Event Diagram (PCED) is adopted here and modified to provide the functional modelling of Internet-based control systems. The PCED was firstly introduced by Yang et al. (2001) as a means to symbolically represent the flow of information between a process, actuating, measuring, and controlling devices in addition to operators. The original PCED illustrates the interaction between six different layers (Lay) of a controlled process, from top to bottom: *Operator* (Op), *Human Input Device/Display* (HID), *Communication* (Com), *Computer* (Comp), *Sensor/Actuator* (S/A), and *Process* (Proc). It is used to represent the exchange of information between these different layers on a qualitative time scale. Assigned to these layers are three different types of nodes (Nod), *object nodes* (Nod_{obj}), *I/O-nodes* ($Nod_{I/O}$), and *computation nodes* (Nod_{comp}): An object node is used to specify that information is retrieved from the process through a sensor, or passed to the process through an actuator, or that information is exchanged between the operator and the human input device (HID). The I/O-nodes mark the exchange of signals between the controller (*i.e.* the computer layer) and the communication layer or the sensor/actuator level. Finally, the computation nodes represent an action of the controller: This is either a jump to another computation node or the evaluation of the controller logic in order to compute a control signal that is delivered to an actuator. The arrowheads specify the direction of the signal flow that is denoted by arcs (Edg) that connect two nodes. The sequence of the nodes in the horizontal direction (from left to right) corresponds to the temporal order in which information is processed. The detail can be found in Treseler et al. (2001).

Compared with the original PCED described above, the PCED shown in Fig. 2.4 has replaced the communication level with an Internet level. Therefore,

2.3 Functional Modelling of Internet-based Control Systems

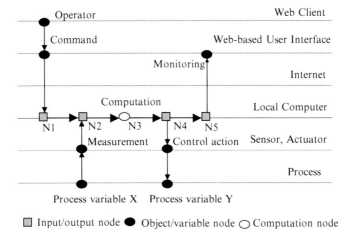

Fig. 2.4 Functional model of Internet-based control systems

the modified PCED is designed for Internet-based control systems. If a certain signal is assumed to be emitted from the *Process* layer, the modified PCED represents the behaviour of Internet-based control systems as the flow of data and information through the layers, *i.e.* as the {B, I, C} triplet.

The *PCED* can be formally defined as a triplet:

$$PCED = (Lay, Nod, Edg) \qquad (2.1)$$

in which *Lay* denotes the six different layers. *INT* denotes the Internet layer:

$$Lay = (Op, HID, INT, Comp, S/A, Proc) \qquad (2.2)$$

Nod denotes the three types of nodes:

$$Nod = (Nod_{obj}, Nod_{I/O}, Nod_{comp}) \qquad (2.3)$$

Edg denotes a finite set of edges, each of which connects two nodes with data transferred as a label:

$$Edg = \{e_1, \cdots, e_n\} \qquad (2.4)$$

The above formal description includes all the information in the {B, I, C} triplet functional model as follows:

$$\begin{cases} B = \{Nod_{comp}^1, \cdots, Nod_{comp}^n\} \\ I = \{Nod_{I/O}^1, \cdots, Nod_{I/O}^k\} \\ C = Edg \end{cases} \qquad (2.5)$$

The advantages gained from using the PCED here are that the structure of the PCED is easy to understand and can be further extended into an information architecture that gives an indication of the physical control system architecture.

2.4 Information Hierarchy

The local computer level in Fig. 2.4 can be further separated into four levels above the sensor/actuator level as shown in Fig. 2.5. These layers are distinguished from each other by "4Rs" principal criteria (Rathwell and Ing 2000):

- Response time: As one moves higher in the information hierarchy, the time delay, which can be tolerated in receiving the data, increases. For example, at the regulatory (control loop) level, data become "stale" very quickly. Conversely, information used at the management and scheduling level can be several days old without impacting its usefulness.
- Resolution: Abstraction levels for data vary among all the levels in the architecture. The higher the level is, the more abstract are the data.
- Reliability: Just as communication response time must decrease as one descends through the levels of the information architecture, the required level of reliability increases. For instance, host computers at the management and scheduling level can safely be shut down for hours or even days, with relatively minor consequences. If the network, which connects controllers at the supervisory control level and/or the regulatory control level, fails for a few minutes, a plant shutdown may be necessary.
- Reparability: The reparability considers the ease with which control and computing devices can be maintained.

As shown in Fig. 2.5, the Internet can be linked with the local computer system at any level in the information architecture, or even at the sensor/actuator level.

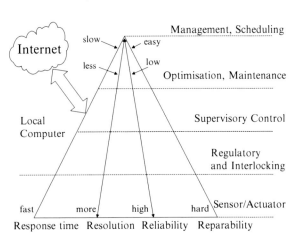

Fig. 2.5 Information hierarchy

2.4 Information Hierarchy

Table 2.1 Pros and cons of possible links between the Internet control levels and existing control levels

Existing information level	Information exchange	Advantages	Disadvantages
Management level	Commercial data systems	Transmits commercial data to customers and managers effectively	Not suitable for real-time monitoring and control tasks
Plant-wide optimization level	Global database	Effective access to plant-wide process information	Not suitable for real-time monitoring and control tasks
Supervisory level	Process database	Easy access to the real-time status of process plants, suitable for implementing advanced control	Missing management information
Regulatory level	PLC, control unit	Allows controllers to directly talk to the Internet	Introduces a high risk of being attacked by malicious hackers
Sensor/actuator level	Smart devices	Monitors and controls the smart devices directly from the Internet	Introduces a high risk of being attacked by malicious hackers

These links result in a range of 4Rs (response time, resolution, reliability, and reparability) depending upon the level at which the connection is made. For example, if a fast response time is required a link to the control loop level should be made. If only management information is needed, the Internet should be linked with a higher level in the information architecture such as the management level or the optimization level.

Table 2.1 shows a simple evaluation for the possible links. This table can be used to guide the selection of the links between the Internet and the existing levels in the information hierarchy. For example, as shown in Table 2.1, smart devices (Internet-enabled sensors and/or actuators) can directly communicate with the Internet with their Internet capability. Programmable Logic Controllers (PLCs) can be directly integrated with the Internet using a Transmission Control Protocol/Internet Protocol (TCP/IP) card (Maciel and Ritter 1998), which allows them to talk to the Internet. However, in most cases direct access to a sensor, an actuator, or a controller introduces a high risk of being attacked by malicious hackers and is probably not desirable. Furthermore, information exchange between process plants and Internet-based clients can be achieved through corporate systems – such as commercial data systems, relational databases or real-time databases, instead of control units. For example, information from the corporate system can be wrapped in a self-describing object written in the Java programming language, seamlessly and efficiently sent to the client's workstation, ready to be published, or included in a form suitable for use in most application software. The disadvantage of these high-level links is the loss of real-time data from the process plants.

2.5 Possible Implementation of Information Architecture

Various system monitoring and control requirements are fulfilled by the links with the Internet established at the different levels in the information architecture. Considering the case that links with the Internet at all the information levels are essential in order to meet some requirements, a possible implementation would be the integrated–distributed architecture as shown in Fig. 2.6 and discussed below.

Nowadays, many control elements have been embedded with Internet-enabled functions, for example, PLC with TCP/IP stack (Maciel and Ritter 1998), smart control valves with a built-in wireless communication based on the TCP/IP, and

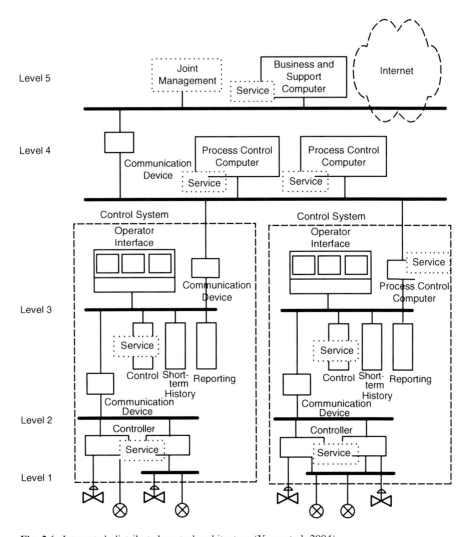

Fig. 2.6 Integrated–distributed control architecture (Yang et al. 2004)

process control computer (DCS) with an Internet gateway. All these Internet-enabled control elements provide services and form a Web-centric distributed architecture for control systems. Furthermore, if some required service is not formally embedded into control elements, it is often practical to facilitate the service by assigning them to a central computational and/or management device. In that case, the service should be located at the same level as its information source. Therefore, a Web-centric distributed architecture of control systems is a reality. Figure 2.6 shows all the services, represented in dotted line boxes, at different information levels in a control system hierarchy.

However, it could prove to be not practical for the system management and users to manage and use these services if too many services are available simultaneously on the Internet and are not integrated properly. The joint management shown in the top level in Fig. 2.6 is a promising way to overcome these problems. The individual services register themselves with the joint management, via a number of physical links, and provide it with a set of parameters describing the service that they can offer. Once the services have been registered, the joint management serves as the repository of all these services.

There are several reasons for doing the management in this way. Firstly, the joint management is the only agent that directly communicates with Web clients in control systems. All control elements are located behind the joint management. Any malicious hackers trying to attack the control systems are actually attacking the joint management rather than the control system itself. This means that the system needs to only defend one access point. This structure will consequently reduce the risk of being attacked by malicious hackers. Secondly, the integrated architecture reduces the number of actual links with the Internet to one, but, at the same time, provides control elements with unlimited virtual links with the Internet. This meets the requirement for linking with the Internet at all the information levels. Finally, most control elements have a very limited capability for networking. The joint management is located at the top level of the control architecture and can offer a great potential for networking for the control system. Therefore, this structure improves the capability of networking and is consistent with congestion control in the Internet.

2.6 Summary

The requirements specification is the first step for the design of Internet-based control systems. It determines the architecture of this new type of control systems. This chapter represents the requirements in a {B, I, C} triplet-based functional model – in this case the functional model is a PCED description, in which the structure information is clearly embedded in it. The information architecture is further derived from the computer level in the PCED and assessed in terms of 4Rs (response time, resolution, reliability, and reparability). The links with the Internet at various levels in the information architecture have been evaluated.

Distributed service and integrated management architecture have been proposed as an ideal implementation, in which the links with the Internet at all the information levels are placed and managed by a joint management agent.

References

Iung, B., Neunreuther, E., and Morel G., (2001) Engineering process of integrated – distributed shop floor architecture based on interoperable field components, *International Journal of Computer Integrated Manufacturing*, 14(3), pp. 246–262.

Levson, N.G., Heimdahl, M.P.E., Hildreth, H., and Reese, J.D., (1994) Requirements Specification for Process-Control Systems, *IEEE Trans. on Software Engineering*, 20(9), pp. 684–707.

Maciel, C.D., and Ritter, C., (1998) TCP/IP Networking in Process Control Plants, *Computers and Industrial Engineering*, 35(3/4), pp. 611–614.

Rathwell, G.A., and Ing, P.E., (2000) Design of enterprise architectures, http://www.pera.net/Levels.html

Treseler, H., Stursberg, O., Chung, P.W.H., and Yang, S.H., (2001) An open software architecture for the verification of industrial controllers', *Journal of Universal Computer Science*, 7(1), pp. 37–53.

Yang, S.H., Stursberg, O., Chung, P.W.H., and Kowalewski, S.. (2001) Automatic safety analysis of computer-controlled plants, *Computers and Chemical Engineering*, 25, pp. 913–922.

Yang, S.H., Chen, X., Edwards, D.W., and Alty, J.L., (2002) Designing Internet-based control systems for process plants, *The 4th Asian Control Conference*, Singapore, September.

Yang, S.H., Chen, X., and Alty, J.L., (2003) Design issues and implementation of Internet-based process control, *Control Engineering Practice*, 11, pp.709–720.

Yang, S.H., Tan, L., and Liu, G.P., (2004) Architecture design for Internet-based control systems, *International Journal of Automation and Computing*, 1(1), pp. 1–9.

Chapter 3
Internet-based Control System Architecture Design

3.1 Introduction

Internet-based and networked control systems represent control systems that communicate with sensors and actuators over a communication medium (wired or wireless) such as the Internet. Control loops that are closed over a communication network, from a simple field bus to the nested Internet, are more and more common as the hardware devices for network and network nodes become cheaper. It makes real-time control over the network acceptable to industry. In recent years, much attention has been paid to the use of real-time networks in a control loop and the networks have been increasingly used as a medium to interconnect different components in large-scale plants and in geographically distributed systems. Typical applications include industrial automation, large-scale power systems, and intelligent transportation systems. The general control structure of Internet-based/networked control systems is shown in Fig. 3.1 (Yang et al. 2005a). A network becomes a component in the closed loop and connects a controller with an actuator and a sensor with a controller. Two communication time delays are included in the closed loop. T_{ca} is the time delay between the controller and the actuator, and T_{sc} is the one between the sensor and the controller.

This chapter starts with the discussion of the traditional tele-operation systems and then introduces the Internet into the control system by using process control as an example. Three typical control structures are described at the end of the chapter.

3.2 Traditional Bilateral Tele-operation Systems

Unlike local control, remote control allows an operator at one location to control an object at some other, perhaps quite distant, location. Remote control was previously called tele-operation. The distance between the controller and the object controlled can vary widely. For example, in the case of nuclear processing, just a few centimetres of lead glass separate the operator from the controlled object; while in the case of subsea tele-operation, several kilometres of water may divide the two.

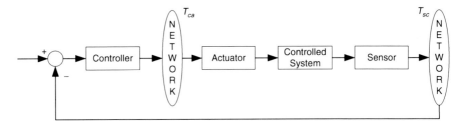

Fig. 3.1 General structure of networked control systems

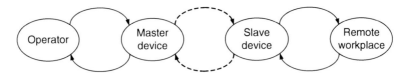

Fig. 3.2 General concept of tele-operation

However, regardless of the application, all remote control systems have the following specific features:

- An operator interface or a master input device that the operator uses to command the system
- A local control device or a slave output device that performs the operator's commanded actions at the remote site
- A communication channel between the remote and local sites

Tele-operation is the performance of remote work. A human operator at the master station controls manipulators to perform work at a remote slave site. The separation between sites may be as simple as a glass window or as complex as a satellite link. The connection between sites may be a mechanical linkage or an electrical signal.

Figure 3.2 shows the general concept of tele-operation. The "world" is divided into two parts: the master site and the slave site. The main aim of tele-operation is that the operator has some task to do at the remote site where they cannot physically be. The tele-operation system extends the operator's capability to be able to work at the remote workplace. This extension is achieved by a master–slave system. The master–slave system is realized as two information channels: action and feedback channels. The action channel, called the feed-forward channel transports the information from the operator to the remote workplace. The feedback channel transports the information to the opposite direction: from the remote workplace to the operator.

The first tele-operation systems were developed by Goertz at the Argonne National Laboratory for use in handling radioactive materials (Sayers 1999). The operator could observe the slave manipulator directly through a thick lead glass window and cause it to move by holding a moving end of the master manipulator.

3.2 Traditional Bilateral Tele-operation Systems

In these earliest implementations, the master manipulator and remote slave arm were exact scaled copies of each other. They were kinetically identical. This guarantees that any singularities or joint limits in the slave manipulator will be in exactly the same place on the master arm. Thus, the operator needs only to operate the slave manipulator.

The two robot arms were also mechanical linked. This provided a natural bilateral system in which any forces applied to one robot arm will be felt on the other. This has the desirable effect that when the operator pushes harder on the master arm, the slave arm pushes harder on the environment. When the environment pushes back, the operator feels that impact on their arm. The negative side effect of realistic force feedback is that the operator actually has to do physical work in order to perform work at the remote site. In later systems, the use of electrically powered manipulators removed the need for a direct mechanical connection.

In the earliest tele-operation systems, the distance between operator and remote site was constrained by the need for the operator to directly view the remote environment. This is called short-range tele-operation. In this case, there is no constraint on the flow of information between sites, and there is no communications delay.

The addition of CCTV camera means that the separation between operators and the remote environments could be greatly increased. These systems enable medium-range tele-operation (or remote control). If the connection between sites in such systems is entirely electrical and based on private media, there will be no perceptible communication delay. The operators may see what is happening via a CCTV camera, they may hear what is happening via a microphone and speaker, and they may feel what is happening via a force-reflecting master manipulator.

In most cases, there will be a communication delay, which means that operators can no longer rely on their reflexes to detect and correct problems that happen at the remote site on the slave device. Instead, they must provide the slave device with sufficient intelligence to enable it to immediately react to problems by itself. Therefore, the communication between the operators and the slave device is only carried out at a more abstract level, rather than at a detailed command level, i.e. raising the level of communication between the operators and the slave device at the remote site.

Intelligent autonomous control (Yang 2005) means embedding part of the operator's knowledge about the use of the remote system at the remote site on the slave device and part at the local site on the master device. The result is that the master device can provide immediate feedback for the operator without needing to wait for the slave device, and the slave device may react immediately to its environment without needing to wait for any operator response. Therefore, the problem caused by the communication delay is significantly overcome with the distributed knowledge more evenly through the system.

The principle of the intelligent autonomous control is shown in Fig. 3.3. At the operator station, the partial copy of the remote site, *i.e.* the model of the slave environment and device, is used to create a virtual reality representation of the slave device at the remote site. This simulation provides the operators with

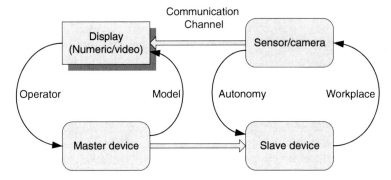

Fig. 3.3 Intelligent autonomous control

immediate sensory feedback. Since the operator interaction is entirely within that simulated world, it is largely insulated from the limitation of the communication delay. At the remote site, the partial copy of the master device is used to create a level of local autonomy that allows the manipulator for the slave device at the remote site to react immediately to sensory data.

The communication between sites is still necessary for supervision and error correction, but at a symbolic level and a lower frequency. This feature permits the remote control tasks to be performed via a delayed low-bandwidth communications link.

To cope with the unexpected event, the system encodes predicted results within the control command stream. As the slave executes each command, it compares expected and actual sensory readings, watching for discrepancies. Small differences may be accommodated by mechanical compliance. Large differences may be the result of an unexpected failure or an inconsistency in the model and the environment. When such an unexpected event occurs, the slave device pauses and sends a signal to the master device. This information may be transmitted from slave to master so as to be immediately available should an error actually occurs. It is then up to the operator located at the local site with the master device to diagnose the error and take corrective action, causing new post-error commands to make their way to the slave device at the remote site.

An example is given in Han's recent work (Han et al. 2001) on a personal robot control. A personal robot simulator and a virtual environment are located at the operator site, equivalent to the master device; an intelligent path generator and a path following controller are at the robot site, equivalent to the slave site in Fig. 3.3. The control architecture is shown in Fig. 3.4. The personal robot simulator and the virtual environment provide the operator with a complete control environment and the controlled object and allow the operator to work within the simulated world. The intelligent path generator restores the moving path of the virtual robot. The path following controller guarantees that the real robot follows the generated path. The path generator and the path following controller can reduce the time difference between the real robot and the virtual robot.

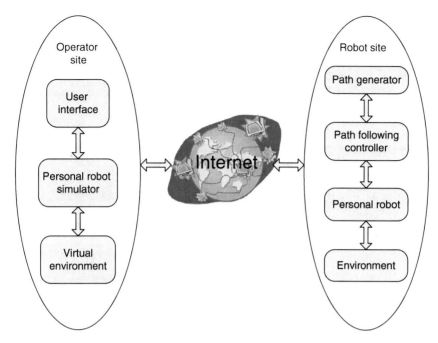

Fig. 3.4 An Internet-based personal robot control

3.3 Remote Control over the Internet

In the last decade, the most successful network developed has been the Internet. It has been widely used as a communication and data transfer mechanism. The Internet can be used as a global platform not only for information retriving but also for remote control. Many Internet applications are being created all over the world. Tele-robotic systems (Han et al. 2001; Baruch and Cox 1996; Golberg and Mascha 1995; Luo and Chen 2000; Paulos and Canny 1996; Taylor and Dalton 2000; Yang et al. 2004a) and virtual laboratories for distance educational purpose (Copinga et al. 2000; Yang and Alty 2002) are two typical Internet-based real-time applications. In the area of control systems, some work has been developed for guiding the design process, dealing with Internet latency, and ensuring safety and security (Yang et al. 2003, 2004b, 2005b).

With the fast development and major use of the Internet, a global information platform has been created for control engineers, allowing them to do the following:

- Monitor and condition of machinery via the Internet
- Remotely control machine
- Collaborate with skilled operators situated in geographically diverse location
- Integrate client needs in production lines

- Manufacture on demand through the Internet
- Provide students in distance learning locations with experimental environments through real and virtual laboratories

Introducing the Internet into control systems brought the benefits described above to industry and also introduced many new challenges to control system designers. These challenges are summarized as follows:

- System architecture design, *i.e.* how to structure this new type of control system, and where to place the Internet in the control system
- Overcoming Web-related traffic delay, *i.e.* dealing with Internet latency and data loss
- Web-based interface design, *i.e.* how to provide an operator an efficient operating environment
- Concurrent user access, *i.e.* dealing with multiple users operating the system simultaneously
- Web-related safety and security, *i.e.* ensuring the safety and security of remote control and stopping any malicious attacks and mis-operation

These challenges make the design methodology for Internet-based control systems different from those for traditional computer control systems. This chapter focuses on the system architecture design. The rest of the challenges are addressed in the following chapters of this book.

If we consider process control as an example, applications range from standalone computer-based control to local computer network-based control, such as a distributed control system (DCS). Figure 3.5 shows a typical computer-based process control system hierarchy. Process control in Fig. 3.5 is broken into the

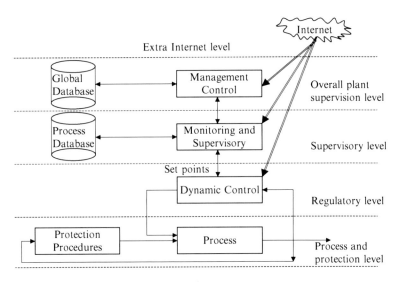

Fig. 3.5 Process control system hierarchy and possible links with the Internet (Yang et al. 2003)

3.3 Remote Control over the Internet

following hierarchical levels: plant-wide optimization, supervisory, regulatory, and protection. The global database and the plant data processing computer system are located at the top level where considerable computing power exists. Process database and supervisory control are located at the second level in which many advanced control functions are implemented. DCS and process and protection are at the lower two levels, respectively. Suppose that the objective of establishing Internet-based process control systems is to enhance rather than replace computer-based process control systems by adding an extra Internet level to the hierarchy. The extra Internet level should be properly placed in the existing process control system hierarchy according to the control requirements. Figure 3.5 shows possible links between the Internet and a process control system. The additional Internet level above the process control system hierarchy might be linked with the existing process control system via the plant-wide optimization level, the supervisory level, or the regulatory level. Therefore, the Internet-based process control is implemented as an Internet control level in the process control hierarchy. Figure 3.6 illustrates the Internet control level as a remote monitoring and control station linked with Field Bus via wireless an Internet connection and a modem. This typical implementation is focusing on remote monitoring and supervision over the Internet rather than real-time remote control.

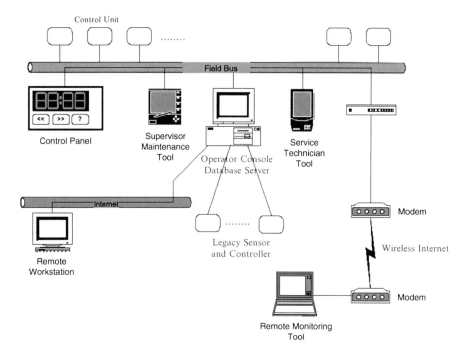

Fig. 3.6 DCS linked with the Internet (Yang et al. 2003)

3.4 Canonical Internet-based Control System Structures

A traditional control system structure is shown in Fig. 3.7. An operator gives a desired reference input to the controller. The controller outputs the control signal to the actuator based on the difference between the desired reference input and the measured output. The actuator passes the control action to the plant. The measured output is fed back to the controller through the sensor.

The straightforward control structure over the Internet is to allow the operator located in the remote site to send control commands through the Internet, such as sending desired input to the controller located with the plant in the local site. The structure is shown in Fig. 3.8. In order to monitor both the performance of the controller and the situation of the plant, the measured output and/or some visual information are required to feedback to the operator at the remote site. Because the Internet is excluded from the closed loop and the controller is located at the same location as the plant, the Internet transmission delay will not affect the performance of the control system. Obviously, the Internet transmission delay will affect the transfer of the desired input from the operator site to the controller site. Some measures must take place to compensate these effects.

In some cases such as a virtual control laboratory and remote design of controllers, it is necessary to locate the controller in the remote site, which is connected with the actuator and the sensor through the Internet, as shown in Fig. 3.9. The Internet has become part of the control system in this case. The transmission delay is introduced in both the actuator and sensor communication channels. These transmission delays will significantly affect the performance and stability of this type of control system. In this structure, the Internet sits between the controller and the sensors/actuators/plant. The sensors/actuators/plant component, which provides the primary data and receives the commands, could be identified as the passive control part; in contrast, the controller, which issues the commands depending on the received data, could be identified as the active control part.

Many existing Internet-based control systems adopt a bilateral control structure, *i.e.* one controller located in the plant site, another in the operator site, and linked through the Internet as shown in Fig. 3.10. For example, based on this control structure, robotic tele-operation uses the controller in the plant site to control the slave device and the one in the operator site to control the master device. Another example is the advanced control for manufacturing processes. Usually, the

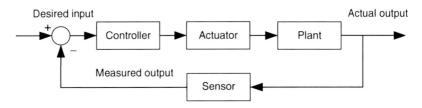

Fig. 3.7 Traditional control system

3.4 Canonical Internet-based Control System Structures 25

Fig. 3.8 Control structure with the operator located remotely

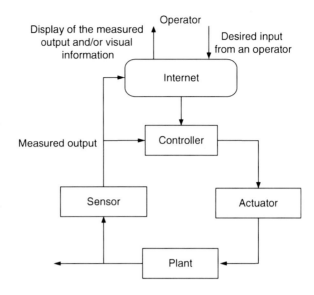

Fig. 3.9 Control structure with the controller located remotely

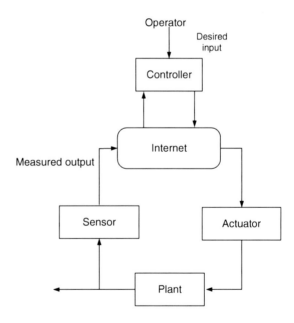

controller in the plant site is responsible for regulating the normal situation. Once the performance of the controller is degraded due to a disturbance in the environment or the change of the production situation, the controller in the operator site is brought into use for tuning the parameters and/or changing the desired input for the controller in the plant site.

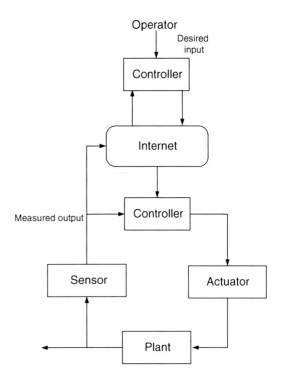

Fig. 3.10 Control structure with bilateral controllers

3.5 Summary

Internet-based control system architecture design is the design task immediately after the requirement specification in the Internet-based control system design. Various requirements might need to be met by different system architectures. For example, in process control Internet-based control aims to enhance rather than replace computer-based control system by adding an extra Internet level to the control hierarchy. This chapter introduces three canonical Internet-based control system structures, with the operator located remotely, the controller located remotely, or with bilateral controllers, *i.e.* one in a local site and another in a remote site.

References

Baruch, J.E.F., and Cox, M.J., (1996) Remote control and robots: an Internet solution. IET Computing and Control Engineering Journal, 7(1), pp. 39–44.
Copinga, G.J.C., Verhaegen, M.H.G., and Van De Ven, M.J.J.M., (2000) Toward a Web-based study support environment for teaching automatic control. IEEE Control Systems Magazine, 20(4), pp. 8.

References

Golberg, K., and Mascha, M., (1995) Desktop tele-operation via the World Wide Web. Proc. IEEE Int. Conf. Robotics and Automation, Nagoya, Japan, May.

Han, K.H., Kim, S., Kim, Y.J., and Kim, J.H., (2001) Internet control architecture for Internet-based personal robot. Autonomous Robots, 10, pp. 135–147.

Luo, R.C., and Chen, T.M., (2000) Development of a multibehavior-based mobile robot for remote supervisory control through the Internet. IEEE Transactions on Mechatronics, 5(4), pp. 376–385.

Paulos, E., and Canny, J., (1996) Delivering real reality to the world wide web via tele-robotics. *Proc. IEEE Int. Conf. Robotics and Automation*, Minneapolis, MN, April, pp. 1694–1699.

Sayers, C., (1999) Remote Control Robotics, Springer: New York, pp. 27–40.

Taylor, K., and Dalton, B., (2000) Internet robots: a new robotics niche. *IEEE Robotics & Automation Magazine*, March, pp. 27–34.

Yang, S.H., (2005) Remote control and condition monitoring, in Chapter 8 of the book *E-manufacturing: Fundamentals and Applications*, Cheng, K. (ed.), WIT Press: Southampton, pp. 195–230.

Yang, S.H., and Alty, J.L., (2002) Development of a distributed simulator for control experiments through the Internet. Future Generation Computer Systems, 18(5), pp. 595–611.

Yang, S.H., Chen, X., and Alty, J.L., (2003) Design issues and implementation of Internet-based process control. Control Engineering Practice, 11(6), pp. 709–720.

Yang, S.H., Zuo, X., and Yang, L., (2004) Controlling an internet-enabled arm robot in an open control laboratory. Assembly Automation, 24(3), pp. 280–288.

Yang, S.H., Tan, L., and Liu, G.P., (2004) Architecture design for internet-based control systems. International Journal of Automation and Computing, 1, pp. 1–9.

Yang, T.C., Yu, H., Fei, M.R., and Li, L.X., (2005) Networked control systems: a historical review and current research topic. Measurement and Control, 38(1), pp. 12–16.

Yang, S.H., Chen, X, Tan, L., and Yang, L., (2005) Time delay and data loss compensation for Internet-based process control systems. Transactions of the Institute of Measurement and Control, 27(2), pp. 103–118.

Chapter 4
Web-based User Interface Design

4.1 Features of Web-based User Interface

Advances in control and information technology have shifted the operator role from being the key element in the control loop to the new function of plant supervisor and troubleshooter. Internet-based process control facilitates this shift since many routine control functions have been taken over by computer-based control system at the regulatory level in the process control hierarchy. The Web-based user interface should be designed to suit this shift. The central design objective for a Web-based user interface in Internet-based process control is to enable the operator to appreciate more rapidly what is happening in process plants and to provide a more stimulating problem-solving environment outside the central control room. It should be borne in mind that media available in the Internet environment outside the central control room will be very much limited compared to those in the central control room. For example, a large-scale screen has been applied to many types of industrial plants (power plants, railway, traffic control system, chemical plants, and so on). The user interface in the central control room has some of the following features:

- Has a wider display area and gives more information
- Has a video wall system to help operators to monitor information from all plants under control, and to share information between operators
- Apply multimedia technology or use various types of displays

However, the Web-based user interface outside the central control room cannot always possess the above features and it may be a simple laptop. This chapter clarifies the steps involved in Web-based user interface design, particularly selecting the most suitable media to display the information.

4.2 Multimedia User Interface Design

The technologies from the areas of "multimedia" and "virtual reality" show considerable potential for improving yet further the human–computer interfaces used in process control technology (Hori and Shimizu 1999), and different media can transmit certain

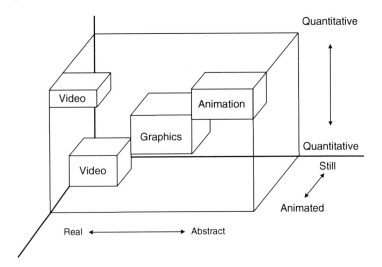

Fig. 4.1 Features of media in interface design

types of information more effectively than others and hence, if carefully chosen, can improve operator performance (Alty 1999). Choosing the best media for different interface tasks and minimizing the amount of irrelevant information in the interface are two main guidelines in the user interface design. This section focuses on choosing the best media for different interface tasks, which are classified into two types of functions (Kawai 1997) according to operational goals: process operation functions and process monitoring functions. Process flowcharts indicate current process status, historical trend displays, and visual information of the plant, which form the elements of process monitoring functions. Operation window displays the elements of process operation functions, including start-up/shutdown button and controller panel, etc.

A medium is an agreed mechanism for communicating information between operators and computer systems. Media have different capabilities for communicating certain types of information. Figure 4.1 shows the features of various media where video, text, graphics, and animation are contrasted using the three axes of qualitative–quantitative, real–abstract, and still–animated. Numbers are more precise than video or graphics. Video is more real than graphics and graphics are more abstract than text. Video and animation are highly animated (Alty 1999).

Multimedia interface is to simultaneously use more than one medium for communicating between human beings and computer systems. For human beings, using multiple media is normal – spoken language, body language, and touch are all used to transmit a message to another person, in parallel. Input and output media are used to communicate with users. Input media are used to communicate information to the system, and output media often involve text, graphics, video, and animation.

The main goals for applying multimedia design techniques in process control interface are as follows:

(a) To enable the operators to appreciate more rapidly what is happening in a dynamic system

(b) To enhance the operator's ability to assimilate what is being presented in a new situation
(c) To provide a more stimulating problem-solving environment

All these situations are concerned with choosing the most appropriate representation to enhance problem solving. The following three properties are essential to all appropriate representations:

(a) All the information required must be currently available in the representation chosen.
(b) The information should be presented in such a way as to be readily perceived and understood in the right context by the user.
(c) Other information, not relevant to the problem-solving process, should be kept to a minimum.

A key question is when should particular media be used, and in what combination, to achieve the operators' goals? It is well known that some media transmit certain kinds of information better than others. The problem has to be approached from two aspects – what do the operators want the information for, and how do the constraints of a particular medium enhance or diminish the capability of the operators to succeed in their goal? We use the following case study to illustrate the answers to the above questions.

4.3 Case Study

This section describes the author's experiences (Yang et al. 2004) in building an interactive Web interface for operating and managing an Internet-enabled arm robot in the author's laboratory. A multimedia technology, Macromedia Flash, has been employed to build a simulator for the arm robot, which is embedded in the client Web interface. Through operating the simulator, the users generate various control commands and send them to the remotely located arm robot without recognizing the difference between the real arm robot and the arm robot simulator. Therefore, the operability of the arm robot has been dramatically improved.

4.3.1 System Architecture

The hardware structure of the Internet-enabled arm robot system is depicted as Fig. 4.2. In the local side, it consists of a physical arm robot, a Web camera, and a server machine. The physical arm robot is connected with the server machine through the RS232 link. The server machine comprises not only the Internet server, but also functions as a local control system for the arm robot. The server is implemented mainly based on Apache HTTP server. The camera is connected to the server in order to provide a visual feedback on the robot status. In the remote

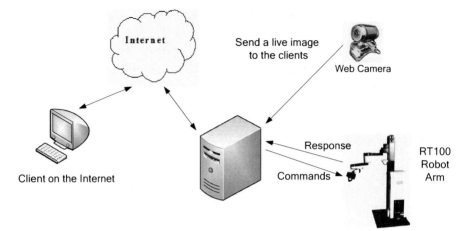

Fig. 4.2 System layout

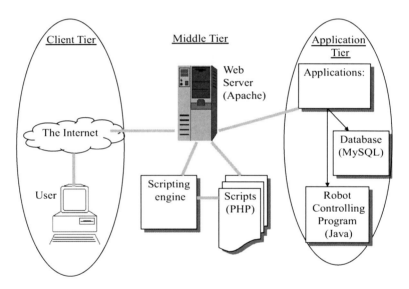

Fig. 4.3 The three-tier architecture

side, Web clients use a 3D interactive interface to operate the arm robot. An arm robot simulator has been employed in the interface to provide a highly flexible and direct way for the robot operation. Macromedia Flash, a multimedia technology, is used in the development of the simulator.

A three-tier architecture is used to implement the above system, as shown in Fig. 4.3. On top is the client tier, in which usually Web browsers are used to interact with the rest of the system. The middle tier is the Apache HTTP server, which is in charge of communication between the other tiers and provides Internet services, such as processing HTTP requests and formulating responses. The PHP scripting language has been chosen as the middle-tier scripting language. PHP is particularly suited for Web database applications because of its integration tools for the Web and database environments. Furthermore, the flexibility of embedding scripts in HTML pages enables easy integration with the client tier. The application tier is at the lowest level, which consists of a mySQL database for user management and a real-time Java program for controlling the RT100 arm robot. The Java program communicates with the Web server through TCP/IP sockets and works independently. The mySQL database is used to log the user's details such as the number of times of logging-in and logging-out and user Web addresses.

4.3.2 Design Principles

The design objective for the Internet-enabled arm robot control is to enable Web clients to directly manipulate the arm robot and appreciate more rapidly what is happening in the physical arm robot located remotely from the users. There are a number of general user interface design principles available (Hori and Shimizu 1999; Alty 1999; Kawai 1997; Sutcliffe et al. 2006), such as user in control, directness, consistency, feedback, and simplicity. A set of design principles of human interfaces for process control based on functional analysis are followed here (Kawai 1997):

Principle 1: Judgement – a system designer should consider the principle that the computer should not replace the judgement functions of operators by providing information on transparency of dynamic process and supporting information of automatic output functions.

Principle 2: Dialogue – a system designer should take into consideration the dialogue function in the interface design by ensuring the consistency of interaction and providing flexible interactive functions.

Principle 3: Monitoring – a system designer should consider the principle that the monitoring function is the basic one in the interface design and should provide global information access functions to the users.

Principle 4: Operation – a system designer should provide a promising environment for the users to make fast responses in interaction and directly manipulate the process.

In applying the above design principles, a task analysis is required to extract the essential information for the operation. The method (Hori and Shimizu 1999) is explained below:

1. Each task is clarified in every operation mode (normal, trouble shooting, start-up/shutdown, etc.).
2. The essential information items for each task are extracted.
3. A display style and media (text, graphics, animation, and video) for each item of information is selected.

A more detailed approach to multimedia interface design was proposed in Alty and Bergan (1995). A task analysis is carried out on individual operation tasks to establish the information requirements. These needs must then be characterized in terms of a defined information processing need. The capabilities that the various media provide are then matched to these needs to obtain a set of possible multimedia candidates. Finally, a desired interface can be derived by checking the possible interfaces through refinements based on lessons learned from user participation and from previous iterations.

4.3.3 Implementation

Macromedia Flash has been chosen for the development of the interface in this case study. Flash allows users to create full screen animation (flash movie) and interactive graphics. The flash file is saved in a vector-based format, which results in extremely compact files. This feature enables the flash-based animation to be embedded in and quickly downloaded from a Web page.

The objective of this implementation focuses on providing a flexible, direct, and easy-controlled 3D interface for the Internet-enabled arm robot. Following the above general design principles and the method of task analysis, we first identify the operation tasks for the arm robot users such as initializing, start, stop, and moving the arm to a desired position, and then match the information for each task with the display style and media. The interface is illustrated in Fig. 4.4. The four principles have been taken into consideration in the interface design. For example, we choose dynamic text to display accurate position of the arm with an animation to show the movement of the arm from an initial position to a desired position. Therefore, the interface provides information on transparency of dynamic process for judgement. Graphics are used to display the operation buttons, in which different working modes are shown in different colours. In doing so, the interface provides a promising environment for the operators to make fast responses in interaction and directly manipulate the arm robot. The Web camera is used to show the actual position of the arm robot in order to get direct access to its real-time information for the monitoring purpose. In order to display the visual feedback from the Web camera, the NetMeeting ActiveX control has been added to the client interface. The visual window is popped out/off by pressing the "OpenCam" button in the interface. A dynamic simulator is embedded in the interface for virtually operating the physical arm robot. With this simulator, users can control the arm robot offline, choose the best position for the arm, and then send the automatically generated control command to the physical arm robot by pressing the "Send data" button. The arm robot will follow the control command to move to the desired position. The embedded dynamic simulator provides a flexible interactive function for dialogue, monitoring, and operation.

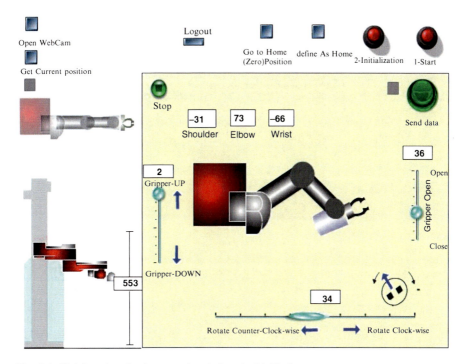

Fig. 4.4 Web interface for the arm robot designed with Flash

4.4 Summary

This chapter discusses the Web-based user interface design for Internet-based control systems. Since there is very limited resource available for remote operation, the Web-based user interface must adopt the most suitable media selected from video, text, graphics, and animation to display the information. Failing to do so will slow down the response in the remote operation. The general design principles are illustrated by a case study for an Internet-enabled arm robot control system. We briefly mentioned another important issue on identifying the essential information items to display in the Web-based user interface. A formal method-based research on this issue can be found in Hussak and Yang (2007) and a detailed discussion is omitted here to avoid shifting away from the core topics in the subject.

References

Alty, J.L., (1999) Multimedia and process control interfaces: signals or noise? *Transaction of the Institute of Measurement and Control,* 21(4/5), pp. 181–190.

Alty, J.L. and Bergan, M., (1995) Multi-media interfaces for process control: matching media to tasks, *Control Engineering Practice,* 3(2), pp. 241–248.

Hori, S. and Shimizu, Y., (1999) Designing methods of human interface for supervisory control systems, *Control Engineering Practice*, 7, pp. 1413–1419.

Hussak, W. and Yang, S.H., (2007) Formal reduction of interfaces to large-scale process control systems, *International Journal of Automation and Computing*, 4(4), pp. 413–421.

Kawai, K., (1997) An intelligent multimedia human interface for highly automated combined-cycle plants, *Control Engineering Practice*, 5(3), pp. 401–406.

Sutcliffe, A.G., Kurniawan, S., and Shin, J.E., (2006) A method and advisor tool for multimedia user interface design, *International Journal of Human-Computer Studies*, 64, pp. 375–392.

Yang, S., Zuo, X., and Yang, L., (2004) Controlling an Internet-enabled arm robot in an open control laboratory, *Assembly Automation*, 24(3), pp. 280–288.

Chapter 5
Real-time Data Transfer over the Internet

5.1 Real-time Data Processing

A basic requirement in any successful application of a web-based system, such as Internet-based control systems, is the provision of efficient real-time processing and data transfer over the Internet. In a significant number of real-world environments, real-time web-based systems involve the transfer and exchange of large amounts of numerical data over the Internet. The heterogeneity and limited traffic resources of the Internet considerably complicate the transfer of such bulky data; for example, if a number of clients simultaneously try to connect with the same server or multiple data sources are accessed over the Internet via different platforms. Such situations require that a data transfer format has to be acceptable to heterogeneous platforms. Additionally, large amounts of data, such as graphics, desktop videos, and images uploaded to the Internet (which may sometimes consist of gigabytes of data), are increasingly being accessed, while, at the same time, the bandwidth available for communication is limited because of the increasing popularity of the Internet.

Most relevant research on real-time data transfer has focused on the data required in high computing applications. Widener et al. (2001) used a number of wrapped message formats, using the eXtensible Mark-up Language (XML) (Brag 2000), to provide flexible run-time metadata definitions that facilitated an efficient binary communication mechanism. Clarke et al. (2001) set up a distributed interactive computing environment, using the eXtensible Data Model and Format (XDMF), which incorporated network distributed global memory, Hierarchical Data Format (HDF) (McGrath 2001), and XML for high performance computing applications. XDMF is an active, common data hub used to pass values and metadata in a standard fashion between application modules and to provide computational engines in a modern computing environment. Nam and Sussman (2003) implemented the HDF format to store National Aeronautics and Space Administration (NASA) remote sensing data within a specific schema. The ScadaOnWeb system (2001) targeted a new standard together with a generic architecture for handling numerical data, in addition to enabling process control, monitoring, and optimization via the web.

5.1.1 Features of Real-time Data Transfer

In Internet-based control systems, real-time data often need to be exchanged between system components, such as sensors, actuator, and controllers, via the Internet. Such real-time data normally have the following features:

- Timeliness: real-time data are time sensitive and have strict time restrictions. The correctness of a system depends not only on the logical correctness of any computations performed, but also on time factors; late data in a stream will result in a media information interruption, while very early data can cause buffer overflow.
- Heterogeneity and complexity: scientific data measure physical phenomena and extend to a large range of data types. Sampling scientific data can be a single binary number, a series of numbers describing physical phenomena, or text and image descriptions of physical devices. A data record may have blocks of many thousands entries, with data corresponding to different times, positions, measuring points, and variables.
- Server Push: real-time applications can be categorized into interactive and streamed applications. Interactive applications are characterized by a two-way exchange of data, examples of which are Internet telephony and distributed simulations. Streamed applications, on the other hand, are essentially one-way flows of information, such as remote monitoring systems. Data exchange in both types of application is pushed by a server, which is responsible for setting up a communication channel, initiating the transmission of data and providing various data access services to remote clients.

Managing and using real-time data in the Internet environment present many challenges, including the following:

- Handling huge volumes of data efficiently: because of the amount of data, it is important to use efficient algorithms and mechanisms for data storage and transfer.
- High data retrieval performance: Internet-based control systems need to offer high-quality, cost-effective data entry services suitable for high volume data computing such as data extraction from a database and accessing and mining data from distributed data sources with minimum operating time.
- Performance across heterogeneous systems: some data are stored and retrieved within the same context; others are shared or distributed to many contexts. Across heterogeneous systems, operating systems are not identical and users are heterogeneous. A storage data context may be incompatible with a particular user's system and may, consequently, need reformatting, translation, or some other form of data modification to create or recreate the intended meaning for the user to understand (Nam and Sussman 2003).

5.1.2 Light and Heavy Data

To simplify real-time processing of heterogeneous data, such data are classified into light and heavy data (the distinction is made in terms of the amount of data, *i.e.* data

5.1 Real-time Data Processing

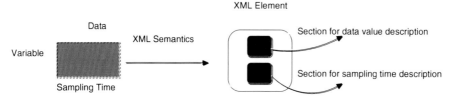

Fig. 5.1 XML data structure

entries, and its physical meaning). Thus, light data are considered to be small quantities of scientific data, while heavy data are considered to be large amounts of data. As a rule of thumb, if data have "more than 300" entries, they are classified as heavy data.

XML is a natural candidate for light data description and binding. An XML document is composed of a hierarchy of elements and can be nested to any level of depth. Elements are logical units of information in an XML document. Real-time data are two-dimensional and consist of a sampling time and variable values. The description of such real-time data in XML is achieved using XML elements. Each data point corresponds to an element. An element has two sections that define its value and sampling time. Information between an element's two tags indicates data content and sampling time, respectively. The structure of real-time data stored in XML is shown in Fig. 5.1. The XML tags are user-defined and application-specific. Light data are wrapped using XML in this chapter. An example is given in Table 5.2

Light data wrapped using XML semantics can be manipulated in Java. There are three popular methods to manipulate an XML document using Java: the Java Document Object Model (JDOM), the Document Object Model (DOM), and the Simple API for XML (SAX). The tree-shaped DOM/JDOM model is suitable for data storage and manipulating an XML document. It is because that data with multiple-level attributes are arranged into a single XML element using a SAX-based method; generated classes under the DOM/JDOM-based method are lightweight; and data binding applications use a minimum amount of memory and can, therefore, run efficiently.

We store heavy data in HDF format using HDF datasets and groups. Data stored in HDF are organized in a hierarchical pattern with two primary structures: groups and datasets. A grouping structure contains instances of zero or more groups or datasets. Metadata contains a group name and a list of group attributes. A dataset in HDF, such as a multidimensional array of numbers, has additional metadata logically associated with it. This metadata describes attributes of the dataset such as the rank of the array and the number of elements in each dimension. Several datasets can form a group. HDF groups and links between the groups can be designed to reflect the nature and/or intended use of stored data. Real-time data in an HDF dataset are shown in Fig. 5.2 as a three-dimensional unit, which consists of the sampling time, the position of the variable in the dataset, and the variable value. The description of variables includes the properties of the data point and an associated explanation of these properties. As shown in Fig. 5.2, all original datasets are first presented in HID and then wrapped with XML, which is described in Sect. 5.2. An example is given in Table 5.3

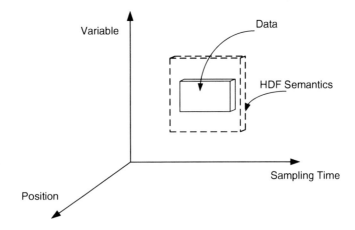

Fig. 5.2 HDF data storage structure

5.2 Data Wrapped with XML

XML provides a standard way to store and structure specific application data and is omnipresent on the Web. Wrapping HDF using XML can fully incorporate all XML features and extend HDF from a local environment to a web environment, thereby making the HDF format acceptable for Internet-based control systems.

Wrapping HDF data using XML consists of two stages: structure mapping and data mapping. Structure mapping is used to ensure that relationships between data described in HDF can be fully expressed in XML. Data mapping is the transference of HDF data content, attributes, and types to an XML format.

5.2.1 Structure Mapping

XML descriptions have a tree structure with a single root and objects nested below their parents. Every JDOM document must contain one root element. Each element can contain as many children elements as needed to represent data in an XML document. JDOM provides Java-specific XML functionality. With the support of JDOM, an HDF group can be mapped to a JDOM document object and its dataset can be mapped to a JDOM element object. There are two cases for wrapping HDF in XML.

Case 1: HDF with a tree structure.
For a simple HDF file, which has a tree structure and is implemented as a directed graph, the HDF file can fully map to XML; in this case, the HDF group is treated as a single JDOM Document object. All objects within an HDF group are treated as

5.2 Data Wrapped with XML

elements of the structure. Attributes of groups can be presented naturally as JDOM Document attributes.

Case 2: HDF with a complicated structure.

For complicated HDF files, which have elaborate grouping structures where some datasets are shared by a number of other groups and have more than one parent, the relationship between groups is not parallel but parental. Therefore, their structure does not match an XML tree structure, as there might be more than one parent for a single dataset. To map these complicated HDF structures to the structure of XML, the structure of HDF with shared datasets needs to be divided into a number of subsections, which divides the overall structure into a number of tree structures. A typical example is shown in Fig. 5.3. Here, dataset 2 is shared by Group A and Group B, and Group A is a member of Group B. To simplify this complicated structure, the structure in Fig. 5.3a is decoupled, as shown in Fig. 5.3b. Mapping the decoupled structure into an XML structure will lead to the XML tree structure shown in Fig. 5.3c. Group A and Group B are mapped as JDOM Document A and Document B, which are at the upper level in the hierarchical structure. Datasets are mapped as JDOM elements. Due to the child–parent relationship between Group A and Group B, Group A is mapped as Element A, a child of Document B. Dataset 1

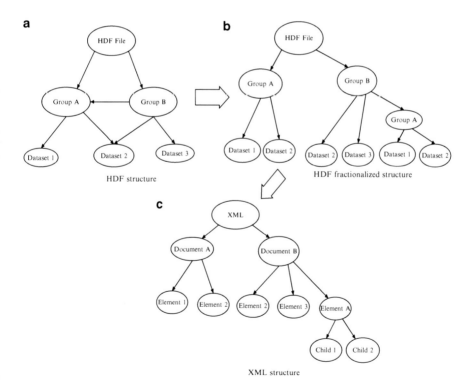

Fig. 5.3 Converting HDF structure to XML structure

and Dataset 2 are mapped as Child 1 and Child 2. The contents of Element 1 and Element 2 in Document A are identical in JDOM to Child 1 and Child 2 in Element A, respectively.

5.2.2 Data Mapping

Data mapping between HDF and XML is achieved via JDOM elements, which have three sections: "DataValue", "DataType", and "DataAttributes". The dataset name, the size of a dataset, and other data attributes of the dataset are categorized as "Data Properties" in HDF and stored in "DataAttributes" of a JDOM element. "DataValue" and "DataType" are identical in both HDF and JDOM. Both JDOM and HDF can be accessed in the same Java environment; therefore, data in HDF can be transferred into JDOM by reading data from an HDF dataset object and writing these data into a JDOM element. These JDOM Document/Element objects are used in the same way as any other Java objects and can be transferred over the Internet in the Java environment via the RMI infrastructure described in the following section.

5.3 Real-time Data Transfer Mechanism

5.3.1 RMI-based Data Transfer Structure

Figure 5.4 depicts the structure of the data transfer mechanism proposed and is built on the RMI infrastructure (Sun 1999). The structure is composed of four basic elements: an RMI server, an RMI Object Registry, a data processor, and remote clients. The RMI server provides back-end communication services. The Object Registry plays the role of object management and provides the naming service for remote objects. The RMI transfer system can pass data objects as arguments and return values, using the full power of object-oriented technology in a distributed computing system. The data processor's role is to collect and process data from actual sites. Owing to its heterogeneous features, transfer data are first categorized into light or heavy data, based on their physical meaning and requirements. Heavy data are organized into HDF format. Both categories of data are then wrapped using XML and processed as JDOM Document/Element objects in the Java environment.

It is necessary to generate a client stub and a server skeleton for RMI communication. A skeleton is a server-side entity that contains a method to dispatch calls to a remote object implementation. A stub is a proxy for a remote object at the client side, which is used to forward method invocations to remote objects.

5.3 Real-time Data Transfer Mechanism

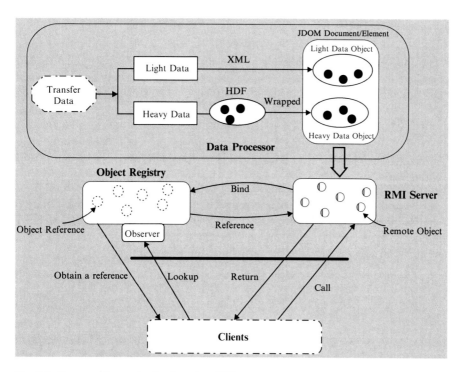

Fig. 5.4 Data transfer mechanism based on RMI

The stub/skeleton will be bound to the client/server through a binding action before objects begin to transfer within the RMI infrastructure. In the RMI infrastructure, Object Registry works together with the RMI server to achieve object transfer. The RMI server creates remote objects and binds each instance with a name in the Object Registry. When a client invokes a method for a remote object, it first looks up the remote object in the Registry using the remote object name. If the remote object exists, the Registry will return a reference for the remote object to the client, which can then be used to make method invocations upon it. As long as there is a reference to the remote object, the object will be reachable from the remote clients.

The sequence of data transfer in the RMI infrastructure is summarized in the following five stages:

Stage 1. Categorize heterogeneous data into light and heavy data according to its features.
Stage 2. Organize heavy data into HDF format and wrap both types of data using XML.
Stage 3. Generate a stub and skeleton for remote objects.
Stage 4. Start the Object Registry and bind remote objects in the Object Registry.
Stage 5. Transfer data by employing the RMI method at every sample interval, with a data object as an argument.

5.3.2 Data Object Priority

Light and heavy data are stored in XML documents, which are treated as data objects in the Java environment. These data objects describe both data values and their structure. By assigning different priorities to data objects, some data objects can be transferred faster than others, so as to meet the different requirements of clients. Normally, vital data are assigned a high priority, and ordinary data given a lower priority. Data with high priority are sent out before data with a lower priority. Priority levels are maintained in the server and updated upon the arrival of any new data transfer requests or when a transmission is completed. Clients can terminate data transfer if necessary.

Data priority submission is implemented using the Java interface PrioritySubmission, in the RMI infrastructure. By extending the interface java.rmi.Remote, the PrioritySubmission interface makes itself available from any virtual machine. The signature of the interface PrioritySubmission is as follows:

```
public interface PrioritySubmission extends Remote {
    public void DataPriority (String [] sequence) throws RemoteException;
    public void EmergencyStop (boolean [] stop) throws RemoteException;
}
```

The DataPriority() and EmergencyStop() method are supported by both the server and client side. The DataPriority() method is used by a client to submit defined data priority to the server. The EmergencyStop() method is used to notify the server to suspend some data transfer. Both methods are defined as being capable of throwing a Java.rmi.RemoteException. The data object name with highest priority is placed at the start of a priority sequence, while less important data objects are placed at the end. The stop sequence is a Boolean array, and defines a logic value True/False, with reference to continuing or terminating data transfer.

Data transfer using various priorities is realized via the proposed observer at the server side, as shown in Fig. 5.4. The proposed observer acts as an object manger. When a client submits an invocation to a remote object, there will be an order to enter the Registry to look up the remote object and obtain references at the server side. The observer is designed to make the order identical to the data priority sequence submitted by the client. In detail, the observer unbinds all remote objects in the Registry before performing a remote object look up. When the invocation is requested from the client, the observer will rebind the remote objects step by step, according to the data priority sequence. The highest priority object is rebound first, while the lowest priority object is rebound last. Only rebound remote objects can obtain a reference from the Registry so that they can subsequently be accessed by the client; other objects, which fail to obtain a reference, continue to wait until they obtain a reference from the Registry. Therefore, by controlling access to a reference, remote object transfer is governed using data priority levels. In the case of a suspension request from a client, the observer will be notified to refuse rebinding of remote objects, which is requested to terminate data transfer, so that references for remote objects can no longer be obtained and data transfer is terminated.

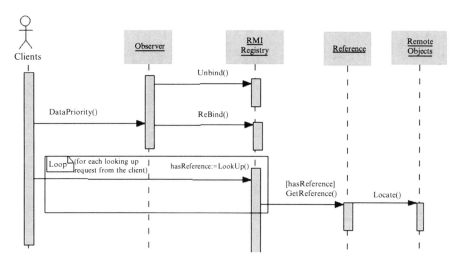

Fig. 5.5 Procedure for looking up remote objects

The procedure of looking up remote objects is shown in Fig. 5.5. Each request from a client repeatedly makes the look up appeal until a reference is obtained. The LookUp() method is responsible for looking up a remote object's reference. The value of the Boolean variable "hasReference" is set by the return value of the LookUp() method. When "hasReference" is true, it indicates that the LookUp() method has found the reference and the method. GetReference() is invocated to obtain the reference and locate the remote object via the Locate() method.

5.4 Case Study

In order to demonstrate the above data transfer mechanism, a remote monitoring system for a reactor process simulator was used as a test bed. The details of its implementation are discussed in this section.

5.4.1 System Description

The reactor process simulator (Yang and Alty 2002) shown in Fig. 5.6 consists of a heat exchanger E201, a reactor R201, and four-hand valves for Nitrogen inlet, liquid outlet, gas outlet, and emergency liquid outlet. The inlet temperature of the reactor is controlled by a PID controller, which manipulates the hot stream flow rate of heat exchanger E201. Figure 5.7 shows the structure of the remote monitoring system, which consists of three parts: the reactor process simulator, an RMI data transfer system, and several remote clients. The local control system

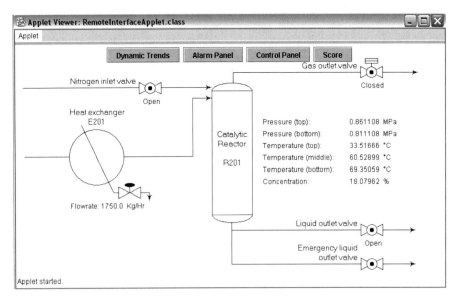

Fig. 5.6 Reactor process simulator (Dai et al. 2006)

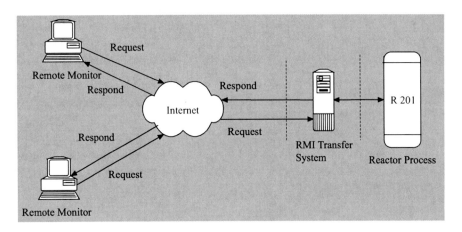

Fig. 5.7 Architecture of the remote monitoring system (Dai et al. 2006)

collects data from the reactor process and passes them to the data processor. The process data are analysed and organized into XML and HDF data formats in the RMI data transfer system. The RMI system is linked to remote clients via the Internet. Remote clients can monitor the local reactor process, define data transfer priority levels, and submit a request to the server. Acquired data are processed as JDOM Document/Element objects and then regularly transferred to the remote clients with a defined priority level (once any requests from clients are accepted).

5.4.2 Priority of Data Transfer

All process variables are categorized as static or dynamic data. Static data include the parameters of the PID controller and the states of the four values (open/close and manual/automatic): nitrogen inlet, liquid outlet, gas outlet, and emergency liquid outlet. Static data are categorized as light data. This type of data is normally of limited size and amount and changes slowly. Dynamic data include all the process variables, *i.e.* reactor temperature, pressure, concentration, and flow rate. It is categorized as heavy data and collected every sampling interval in HDF format.

The main objective of the monitoring system is to remotely monitor control performance; for example how well a controller works, together with a process to track the setpoint of process output. Output variables, such as temperature and pressure in the reactor, the work state of control devices, and PID parameters, are essential for monitoring purposes. Data priority levels are set as shown in Table 5.1 in terms of the monitoring system specification. All light data are set with the highest priority. The small data size of the light data utilizes limited data communication bandwidth and introduces minor extra load on the communication channel. The transfer priority of flow rate is set with the lowest priority here.

5.4.3 Implementation

Two XML files were created to store and transfer light data. One contains PID parameters, while the other contains the work state and mode of the four control valves. The sampling interval of data collection is 1 s. Each data element stored in an XML file is a real-time data point, which includes the value of the variable and its corresponding sampling time. Heavy data are collected in an HDF file; the HDF file has four groups which collect the pressure, temperature,

Table 5.1 Data priority

		Transfer priority rank
Light data	State of nitrogen inlet valve	1
	State of gas outlet valve	
	State of liquid outlet valve	
	State of emergency liquid outlet valve	
	PID parameters	
Heavy Data	Concentration	1
	Pressure (top)	2
	Pressure (bottom)	
	Temperature (top)	3
	Temperature (middle)	
	Temperature (bottom)	
	Flow Rate	4

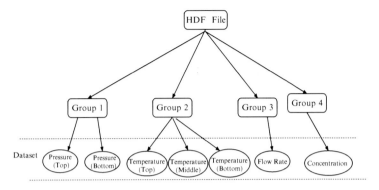

Fig. 5.8 HDF data structure

concentration, and flow rate. The structure of the HDF file is shown in Fig. 5.8. It has a tree structure and can fully match the XML structure; each data variable is three-dimensional: the sampling time, the variable position in the dataset, and the variable value. The data type of the datasets is "Double", and the data length is 300, which covers a 5 min sampling period.

The HDF file is mapped to JDOM data objects in XML. The four groups in the HDF file are converted into four JDOM documents; the datasets are mapped to the elements of the JDOM documents; property information is stored in the element attributes. Mapping to the XML elements, "DataValue" has two XML tags: <Variable value> and <SamplingTime>. The position of data points is not used here. Variable values and sampling times are inserted between these two opening/closing tags. "DataType" is represented using the tag <DataType>, and "DataAttributes" using the tag <Attributes>, and saves the data dimension and size of the amount of the HDF dataset. All data are presented as JDOM objects in the Java environment. Remote clients access the JDOM objects by invoking the LookUp() method. RMI security manager has been installed at both the local and remote side, which allows a legal Java virtual machine to download stub and skeleton class files.

5.4.4 Simulation Results and Analysis

Tables 5.2 and 5.3 illustrate the wrapping of light and heavy data. Using the RMI infrastructure, the work status of the four control valves and the PID controller parameters is wrapped in XML as light data objects and sent to a remote monitoring site via the Internet. User-defined tags include <PID_P>, <PID_I>, <PID_D>, and <SamplingTime>. All PID parameters are wrapped between the opening and closing tags of these tags. Table 5.2 provides an XML file for PID controller parameters. Each data object, such as PID_P in Table 5.2, is treated as an element object in JDOM, which has two sections: the PID_P parameter value and sampling time.

5.4 Case Study

Table 5.2 XML file of light data

```
<Simulation>
 PID.xml <Sector>
            <Parameters>
                    <PID_P>50</ PID_P >
                    <Sampling Time>11:34:47 12/08/2004</ Sampling Time >
                    <PID_I>5</ PID_I >
                    <Sampling Time>11:34:47 12/08/2004</ Sampling Time >
                    <PID_D>0</ PID_D >
                    <Sampling Time>11:34:47 12/08/2004</ Sampling Time >
            </Parameter>
 PID.xml </Sector>
</Simulation>
```

Table 5.3 XML file of heavy data

```
<HDF>
HDF.xml
 <Sector>
 <RootGroup>
      <GroupName>Root</GroupName>
 <Group>
        <GroupName>Group4</GroupName>
        <DataSet>
 <DataSetName>Concentration</DataSetName>
          <HDF5File>"HeavyData.h5" </HDF5File>
         <Section>
       <Attributes>2D % 300</Attributes>
       <DataType>Float</DataType>
 <Concentration>18.10 18.08 18.06 18.02 18.00 17.99
… …   … ….
           </Concentration>
< SamplingTime>19:38:57 19:38:58 19:38:59 19:39:00 19:39:01 19:39:02
… …  …  ….
           </SamplingTime>
          </Section>
        </DataSet>
  …
      </Group>
 …
    </RootGroup>
    </Sector>
   </HDF>
```

Heavy data are collected in the HDF format wrapped as JDOM Document/Elements. Table 5.3 shows the tree structure and individual dataset in an XML file for heavy data, in which a super group "Root", its subgroup "Group4", and its dataset "Concentration" are nested with various XML tags. The heavy data of this dataset are transferred from an HDF file named "HeavyData.h5". The attribute describes the dataset as

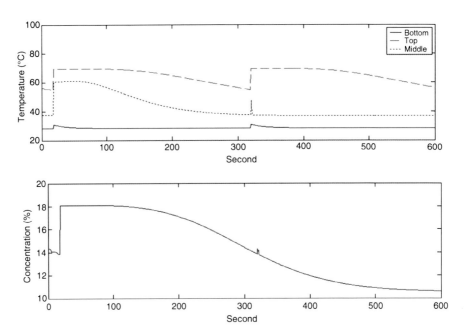

Fig. 5.9 Data received at the remote client side

two-dimensional, with unit "%", type "Float" and 300 data points. Both light and heavy data are stored as JDOM Document/Elements and incorporated into the RMI transfer system. The server binds the data objects in the Object Registry so that any data objects can be looked up from a remote site. The data priority policy is applied in the RMI-based data transfer mechanism. Figure 5.9 shows the concentration of product and temperatures received at the remote client site, at the top, middle, and bottom of R201.

5.4.5 Advantages of RMI-based Data Transfer

Traditional data transfer methods are usually based on socket technology, which is an end-to-end communication mechanism used to establish a connection between a client and a server program. The server communicates with the client via a socket and listens for client connection requests. Communication with a client is conducted by a data stream read from and written to the socket. To assess the proposed RMI-based data transfer scheme, a standard TCP socket-based data transfer method was implemented to carry out the identical data transfer tasks described above.

Two significant advantages were identified as follows:

- The amount of data presented using HDF and XML is smaller than that presented as a data stream when using socket technology to transfer data. Therefore, data transfer latency using the proposed RMI-based method may be shorter than that using traditional socket technology. In a comparison study, a 65,536-byte dataset was collected and transferred to a remote client in 1.553 s. The same dataset when stored in HDF format only required 22,630 bytes, because HDF supports a highly customized data filtering and scatter-gather algorithm (McGrath 2001). After wrapping in XML, the dataset was extended to 39,297 bytes due to the addition of XML tags; however, this is still much smaller than the original dataset. The RMI-based data transfer method took 1.492 s to complete the data transfer, which is slightly faster than the TCP socket method.
- Data objects presented using JDOM are well structured and easy to locate, query, and retrieve. Therefore, the proposed RMI-based data transfer method is more efficient during data query and retrieval than data streams in a traditional socket approach. In detail, all data in the RMI-based data transfer method are wrapped using XML and formed and transferred as JDOM data objects (which are structured as hierarchical trees and can be directly accessed by clients as local objects). To retrieve data, a client only needs to make an invocation on objects. Under the tree structure, data retrieval in terms of data objects is simple and fast for client applications.

5.5 Summary

Internet-based control systems need an effective data transfer mechanism to process and transfer real-time heterogeneous data. This chapter introduces an RMI-based data object transfer mechanism. Heterogeneous transfer data are categorized into two basic groups: light and heavy data, based on physical meaning and data amount. Light data are wrapped in XML. Heavy data are wrapped in HDF initially and then in XML. Both light and heavy data are processed as JDOM data objects in the Java environment. JDOM data objects are transferred over the Internet using the RMI mechanism. This structure is flexible as it not only allows a server to transfer data objects, but also allows clients to define and submit a data priority sequence to the server and therefore to guide data transfer. The main advantages of the mechanism introduced here are as follows: (a) it simplifies heavy data retrieval in client applications via remote invocation of remote data objects and a hierarchy of data objects and (b) the size of a dataset is reduced because of the compact storage structure of HDF; therefore, transmission latency might be reduced. The DMI-based data transfer mechanism has great potential to improve the efficiency of data transmission and therefore meet the requirements of data transfer over the Internet in real-time applications.

References

Brag, T., (2000) The real benefit of XML, Available at: http://www.fcw.com/fcw/articles/2000/0522/tec-bragg-05-22-00.asp.

Clarke, J. A., Hare, J. J., Brown, J. D., (2001) Implementation of a distributed data model and format for high performance computing applications, Available at: http://www.hpcmo.hpc.mil/Htdocs/UGC/UGC00/paper/jerry_clarke_paper.pdf.

Dai, C, Yang, S. H., Knott, R., (2006) Data transfer over the Internet for real-time applications, *International Journal of Automation and Computing*, 4, pp. 414–424.

McGrath, R. E., (2001) HDF5 compound data: technical issues for XML, Java and tools, Available at: http://hdf.ncsa.uiuc.edu/HDF5/XML/tools/compound-data.html.

Nam, B., Sussman, A., (2003) Improving access to multi-dimensional self-describing scientific datasets, Available at: http://csdl.computer.org/comp/proceedings/ccgrid/2003/1919/00/19190172abs.htm.

ScadaonWeb, (2001) Description of work, Available at: http://www.scadaonweb.com/publications/ScadaOnWeb-descriptionofwork.pdf.

Sun Microsystems, (1999) Java™ remote method invocation specification, Available at: http://java.sun.com/j2se/1.5/pdf/rmi-spec-1.5.0.pdf.

Widener, P., Eisenhauer, G., Schwan, K., (2001) Open metadata formats: efficient XML-based communication for high performance computing, *10th IEEE International Symposium on High Performance Distributed Computing*, pp. 315–324.

Yang, S. H., Alty, J. L., (2002) Development of a distributed simulator for control experiments through the Internet, *Future Generation Computer Systems*, 18, pp. 595–611.

Chapter 6
Dealing with Internet Transmission Delay and Data Loss from the Network View

6.1 Requirements of Network Infrastructure for Internet-based Control

Overall, in most cases, Internet-based control systems have been developed by means of extending discrete control systems, which do not explicitly consider Internet transmission features. For example, the laboratory control system developed by Overstreet and Tzes (1999) achieves the Internet-based control by adding the Internet communication elements between the remote controller and the sampling switches, which maintains the discrete control structure. This structure implies that the Internet communication is no more than a bounded time delay element to the control system. This assumption is not true since the Internet transmission is characterized as unpredictable time delay and data loss, which is difficult to model at any period time. In fact, the Internet is a public and shared resource in which various users transmit data via the Internet simultaneously. The route for data transmission from end to end may not be same for every data exchange, and collisions may occur when two or more users transmit data via the same route simultaneously. In order to really achieve Internet-based control system, the underlying mechanism of the ideal network infrastructure should meet the following requirements.

6.1.1 Six Requirements for Ideal Network Infrastructure for Internet-based Control

- Real-time transmission. It should provide the appearance boundary of the transmission time, which can be programmed from API.
- Reasonably reliable transmission. It should provide connection-oriented and retransmission mechanism. During the certain transmission time, it maximizes the effort to guarantee the information delivery.
- Timeout notification mechanism. When a certain time has passed by, it should terminate sending and/or receiving actions and inform the application.

This is vital information for the application to efficiently deal with the time delay and data loss condition.
- Priority-based transmission. The control system communication could involve several catalogues of control signal. Depending on the crucial level for the control system execution, priority levels are assigned to the different catalogue signals. According to the priority, the most crucial control signals should have the highest priority to be transmitted, while those with less priority are transmitted afterwards.
- Time synchronization. Since the control systems are not only logically but also physically distributed around world, it is essential to synchronize the clock of the different part of system, so that the control action can be coordinated.
- Penetrating firewall. Any local network will be protected by one or more firewalls. Internet-based control systems need to penetrate these firewalls and establish a connection between any two nodes on the Internet.

The ideal network infrastructure involves the consideration of Internet transmission behaviours from the control system perspective and requires a new control structure and time delay compensation elements for Internet-based control systems.

6.2 Features of Internet Communication

Different from other private transmission media, the Internet is a public transmission media, which could be used by many end users for different purposes. The major obstacle of Internet-based control systems is how to overcome uncertain transmitting time delay and data loss problems. According to a study of the Internet transmission (Luo and Chen 2000), the performance associated with time delay and data loss shows large temporal and spatial variation. As this study shows, the performance decreases with the distance as well as the number of nodes traversed. The performance also depends on the processing speed and the load of network nodes. Uncertain transmission time delay and data loss problems are not avoidable for any Internet-based application. The reasons why the variable time delay occurs are as follows (Magyar et al. 2001):

- Network traffic changes all the time because multiple users share the same computer network.
- Routes or paths of data transmission decided by Internet Protocol (IP) are not certain. Data are delivered through different paths, gateways, and networks whose distances vary.
- Large data are separated into smaller units such as packets. Moreover, data may also be compressed and extracted before sending and after receiving.
- Using TCP/IP protocols, when error in data transmission occurs, data will be retransmitted until the correct data are received. Therefore, data loss and time delay in data transmission can be treated identically.

In detail, the Internet time delay is characterized by the processing speed of nodes, the load of nodes, the connection bandwidth, the amount of data, the transmission speed, etc. The Internet time delay $T_d(k)$ at the instant k can be described as follows (Han et al. 2001):

$$T_d(k) = \sum_{i=0}^{n}\left[\frac{l_i}{c} + t_i^R + t_i^L(k) + \frac{M}{b_i}\right]$$
$$= \sum_{i=0}^{n}\left(\frac{l_i}{c} + t_i^R + \frac{M}{b_i}\right) + \sum_{i=0}^{n} t_i^L(k) = d_N + d_L(k), \quad (6.1)$$

where l_i is the ith length of the network link, c the speed of light, t_i^R the routing speed of the ith node, $t_i^L(k)$ the delay caused by the ith node's load, M the amount of data, and b_i the bandwidth of the ith link. d_N is a term, which is independent of time, and $d_L(k)$ is a time-dependent term. Because of the time-dependent term $d_L(k)$, it is somewhat unreasonable to model the Internet time delay for accurate prediction at every instance. Internet time delay must be explicitly handled through the network infrastructure, control system architecture, and time delay compensation algorithms.

6.3 Comparison of TCP and UDP

The previous results of the Internet transmission features are normally obtained by using the Internet Control Message Protocol (ICMP), which can avoid the possibility of flooding the network with probe packets. However, it cannot be used as the basis to deliver control signals. Therefore, it is necessary to further investigate the behaviours of the protocols, which can deliver the control signals. Generally, the protocols fall into two categories (Comer 2006): reliable protocol represented by Transmission Control Protocol (TCP) and unreliable protocol represented by User Datagram Protocol (UDP).

As a reliable protocol, TCP is a connection-oriented and full duplex service and guarantees a reliable bidirectional stream of data by means of retransmitting data, discarding duplicating data, and reordering out-of-sequence data packets. As a result, TCP functions poorly in time-sensitive application over long distances with many routing hops. As an unreliable protocol, UDP is a simple and connectionless protocol. It preserves datagram boundaries between the sender and the receiver. Consequently, it could provide real-time transmission. However, there is absolutely no guarantee that the datagram will be delivered to the destination host. Not only can the datagram be undelivered, but also it can be delivered in an incorrect order. Another factor, which affects TCP and UDP transmission, is the presence of a firewall. In order to prevent UDP from flooding the network, UDP traffic is blocked by restrictive firewalls, which eventually cause the failure of the UDP transmission.

Combining with the features of the Internet transmission, the different protocols determine the performance of delivering control signals in some aspects.

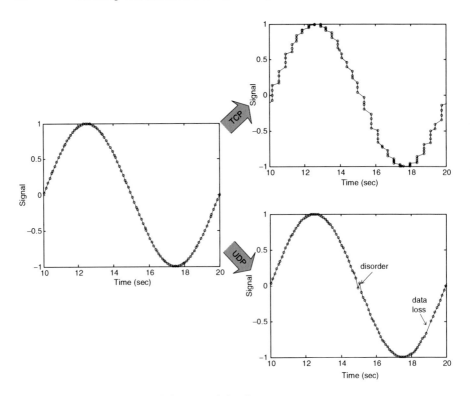

Fig. 6.1 TCP and UDP transmission control signals

As shown in Fig. 6.1, the signal source is designed as the sine function $y(t) = \sin(0.2\pi t)$ and the sampling time is given as 100 ms. Through the TCP transmission, the information of the control signals is compressed at some points, caused by the time delay, so that the function shape is almost lost. TCP ensures the control signals in the same order and the same number. In contrast, through UDP transmission, the function shape is almost maintained, which is caused by the small overhead of the UDP transmission. However, disorder and the loss of the control signals do occur, although irregularly. Another factor shown in the experiment result is that TCP requires extra bandwidth and processing overhead to perform the error corrections. It is clear that neither TCP nor UDP can fully fulfil the requirements of the Internet-based control systems. Therefore, it is necessary to develop a new network service to meet the requirements based on the available technologies.

6.4 Network Infrastructure for Internet-based Control

In order to establish a dedicated network infrastructure to meet the six requirements mentioned above, we organize TCP, UDP, RCP (Real-time Control Protocol) (Perkins 2003), and NTP (Network Time Protocol) (Rybaczyk 2005) together,

6.4 Network Infrastructure for Internet-based Control

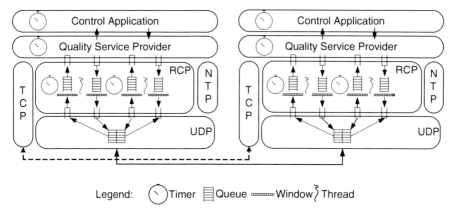

Legend: ⌕ Timer ▤ Queue ═══ Window ⌇ Thread

Fig. 6.2 Network infrastructure for Internet-based control

as shown in Fig. 6.2 (Chen and Yang 2002). The new network infrastructure adjusts the strict layering between the protocols and adopts a more cooperative relationship between them. The structure, which could be considered as an application layer, replaces a complicated protocol TCP with a simple UDP-based framework, which can be used more flexibly and directly. The network service is divided into three layers: UDP, RCP, and quality service provider, with TCP used for initialization and NTP for time synchronization. The communication process starts by using TCP to initialize a connection, which can penetrate firewalls. In detail, while one end opens a TCP server socket associated with a well-known port to accept the incoming connection request, the other end creates a TCP socket and requests a connection. Once this virtual connection is created, the two sides begin to negotiate the transmission specification via TCP. After the completion of the negotiation, the connection is established and kept alive for the entire period. The communication process continues with RCP beginning real-time transmission by binding the underlying UDP to TCP. Afterwards, the communication relies on RCP, and TCP only provides function to detect the absence of members. Through the entire communication process, RCP will maintain the established connection until the connection is destroyed.

6.4.1 Real-time Control Protocol

RCP was mainly designed to address two issues:

(a) Reliable transmission, providing similar functions of TCP to solve the reliability problems such as data loss, disorder, and duplication.
(b) Real-time transmission, providing predictable transferring time and priority-based transmission. RCP also provides a mechanism to maintain a constant connection between communication peers.

6.4.1.1 Reliable Transmission

In order to obtain the reliable transmission, RCP employs a queue window mechanism to cope with data loss, disorder, and duplication. As shown in Fig. 6.2, the sender and the receiver maintain a sending window and a receiving window, respectively.

At the sender side, once a packet has been sent out, it is recorded in the sending window. If the sending window is full, the packet will be kept in the sending queue to wait for transmission. The sending window will not free the record until the remote receiver has acknowledged that the corresponding packet has been received. If the record has not been freed and the retransmission time is due, the sender window retransmits the unacknowledged packet. During this process, a unique sequence number is used to facilitate the acknowledgement.

At the receiver end, the expected sequence number is used to facilitate the receiving packet process. If the received packet sequence number is less than the expected sequence number, which means the packet has been received before, the acknowledgement message will be sent to force the sender end to provide a correct packet. If the number is greater than the expected number, which means if the packet loss occurs, the received packet will be kept in the receiving window, and the receiver sends a message to enforce the sending side to retransmit the missed packet. Once the received packet is the expected one, it will be put in the receiving queue, which finally delivers the message to the application.

In fact, the sequence number and the acknowledgement number are embedded in the packet head. In most cases, the control application is the interactive periodical application, so it is unnecessary to use the specific acknowledgement message to confirm the received packets, which potentially reduces the Internet workload and gives an advantage to the control application.

6.4.1.2 Real-time Transmission

The real-time transmission issues consist of predictability, priority, and exception notification. In practice, "real-time transmission" does not mean that the transmission speed would be increased or that the transmission delay and jitter would be avoided, as this is impossible to be achieved without increasing network capacity. This work is based on a simple hypothesis: the requirements of the Internet transmission capacity can satisfy the ordinary condition and different types of control information having different priorities. The real-time Internet communication means that the higher priority control information has more chance of reaching its destination on time through deliberately scheduling transmission tasks. This is intended to increase the predictability of the Internet transmission. In abnormal situations, any transmission failure can be detected and analysed so that the exception can be efficiently handled. From the designer's perspective, it is primarily concerned with explicitly scheduling retransmitting time, being informed of the sending and receiving timeout, and processing the sending and receiving packets based on the priorities.

6.4 Network Infrastructure for Internet-based Control

In order to address the above concerned issues, two elements have been introduced to RCP: a timer and a connection thread. As shown in Fig. 6.2, each connection has its own timer and connection thread. The main task for the timer is to generate the event that triggers the retransmitting action, determined by the required operation period. When the sending and/or receiving timeout occurs, the timer responds by informing the related parts to handle the exception. In addition, it provides the standard time for the receiver to check the packet timestamp. If the timestamp has expired, the receiver will drop the packet, and at the same time, the receiver informs the sender about the packet being dropping. Since every pair of sender and receiver windows will be operated in an individual connection thread, it enables to process packets in the right order by setting different priority of the thread. It could solve potential priority inversion in cases where a packet with a low priority arrives just before one with a high priority. Furthermore, the scheduling of the packets in the queue depends on the assigned priority of the packet.

6.4.2 Quality Service Provider and Time Synchronization

Although RCP provides the essential service for control applications, it is very complicated and difficult to comprehend. Quality service provider fills the gap between the control application and RCP by means of encapsulating the RCP functions. Furthermore, it also provides some essential services for RCP.

For the control application, it provides an API, which is similar to the well-accepted socket API such as "Connect", "Accept", "Send", and "Receive". It also provides some extended APIs, which relate to real-time features such as "SetPriority", "SetSendTimeout", and "SetReceiveTimeout". It also utilizes the control application data transmission by means of fragmenting the data into the proper size packet assigned the relative priority and the timestamp. At the same time, the control application provides the essential information to the quality service provider, so that it can determine the connection establishment.

For RCP, the quality service provider supports the connection establishment. When the connection request arrives, it includes the expected period of time and priority. It will determine the request acceptance by estimating the capability of network and application. Furthermore, it enables passing exception events to applications.

In addition, the quality service provider provides the time synchronization, which is an essential element for the control application and RCP. In the single process system, it is easy to facilitate with standard timer devices or simply uses the micro-process timer function. In fieldbus systems, it is facilitated by introducing the external timer devices as part of the fieldbus system infrastructure. For the Internet-based control, although the resolution of the clock tick is not required as accurate as at the controller and device levels, it is still an essential part of the system. Currently, most Internet-based control systems use the local time clock instead of a global time clock, which could lead a serious problem. In fact, since the system is available worldwide, the local time clocks, which exist in different time zones, could be out of synchronization and, consequently, are not suitable for an Internet-based control system. This problem can be solved by means of synchronizing the local clock of all the control units and the protocol with a sufficient accuracy time source. NTP server is one of the solutions. According to the survey done by Mills (1997), there are more than 36,000 servers available for the public, and the error of the most NTP clocks is within 21 ms (microsecond) of their servers. There may be additional errors due to asymmetric paths, but this is bounded by one-half the round-trip delay. It could lead to a more accurate result by combining several servers' results together. Overall, the time clock synchronized at a hundred ms level is sufficiently accurate for both the control application layer and the protocol layer.

6.5 Typical Implementation for Internet-based Control

6.5.1 Experimental Set-up

The pilot study uses two water tanks in the process control laboratory at Loughborough University as a test bed. As shown in Fig. 6.3, both of the two water tanks are similar in structure. The inlet flow of the tank is controlled by a hand valve; the outlet flow of the tank is controlled by a pneumatic valve. The control system collects the data from the liquid level sensor via a data acquisition instrument and sends the command to the control valve, so that the liquid level of the tank can be

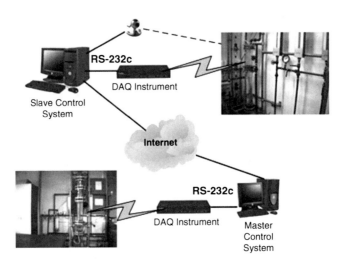

Fig. 6.3 The layout of the experimental set-up

6.5 Typical Implementation for Internet-based Control

kept at a desired value. The control system shown at the top is identified as the slave controller; the one at the bottom is identified as the master controller. The slave one maintains the liquid level at ratio 2:1 of the master. There is a camera at the slave site, which is set up for two purposes. The primary purpose is to test its ability to transmit video data alongside the control data at different priorities, which are 4 and 6, respectively. The maximum and minimum priorities are 10 and 1, respectively. The second purpose is to provide extra information to remotely monitor the slave process. The master and slave control systems are linked via the Internet.

Figure 6.4 presents the control structure of the pilot system in terms of the canonical Internet-based control system structures proposed in Chap. 3. The slave control system receives the desired input from the master control system via the Internet, which is the current liquid level of the master tank. The slave control system adopts the canonical Internet-based control system structure with the operator located remotely as shown in Fig. 3.6. The master control system is a general closed-loop control system linked with the Internet and provides the desired input for the slave control system from a remote site.

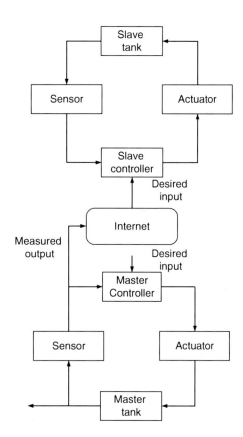

Fig. 6.4 Control structure of the experiment system

6.5.2 Implementation

The canonical Internet-based control system structure shown in Fig. 3.8 has the Internet sitting between the controller and the sensors/actuators/plant. The sensors/actuators/plants component, which provides the primary data and receives the commands, was named as the passive control part; in contrast, the controller, which issues the commands depending on the received data, was named as the active control part. We extend these definitions to the control structure shown in Fig. 6.4 and name the slave control system as the passive part and the master control system as the active part. The Internet sits between the passive and active parts and provides the communication channel for these two parts.

The procedure of implementation of the network infrastructure for the Internet-based control system shown in Fig. 6.4 could be divided into three phases: endpoint initialization, service initialization, and service process. The communication between the passive and active parts is described as follows and illustrated in Fig. 6.5.

- *Endpoint initialization phase.* The passive part creates a server endpoint that is bound to a network address such as an IP address and a port number. The server endpoint listens for connection requests from peers. The "Open" method of the passive part also includes service-specific initialization such as the service capability estimation. For the active part, it uses the "connect" method to actively establish connections.
- *Service initialisation phase.* When a connection request is received by the quality service provider, the provider responds by invoking the "Accept" method. This method performs the strategy for initializing a service handler. It involves assembling the resources necessary to create a new concrete service handle and to assign the connection into this object. The object will be registered to the provider by calling the "RegisterHandler" method. The passive part then calls the "Initial" method to perform service-specific initialization. When the active part successfully establishes the connection, the "RegisterHandler" method will be called to register the active part service handler.
- *Service processing phase.* Once the connection has been established, the passive and active parts enter into the service processing phase. This phase performs application-specific tasks that process the data exchanged between the passive and the active parts. Generally, the passive part initializes the process cycle by calling the "Send" method. The active part receives and processes the data by calling the "Receive" and "Process" methods, respectively. The result is sent back to the passive part. The passive part finally receives the result and carries out control operation. This cycle will be continued until any part quits. During the process, the encoding and decoding of the application data should be performed in the "Send" and "Receive" method, respectively. Once the sending and/or receiving error occurs, while the passive part appends the sending data to the next sending data packet and drops the received data, the active part drops the received data and the sent command.

6.5 Typical Implementation for Internet-based Control

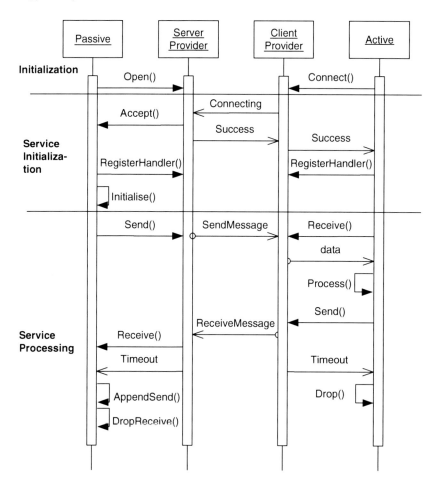

Fig. 6.5 Communication implementation among the passive and active parts

This implementation enhances the reusability, portability, and extensibility of the control systems by classifying the systems into passive and the active modes. It provides the formal structure, which could guide designers to implement the application, enabling designers to use a framework to reduce the implementation effort.

The experimental results are illustrated in Fig. 6.6. The control process achieves the control objective, *i.e.* the slave tank can follow the master's tank liquid level. It should be noted that the value of the master's level has been doubled. As shown in both control process and video frame interval diagram, the Internet congestion did occur during the experiment process, which was represented by video packet being dropped at around 750, 960, and 980 s. However, the control data transmission has not failed during the experiment. Since the control data transmission has a higher priority than the video, it has precedence over the video transmission so that the

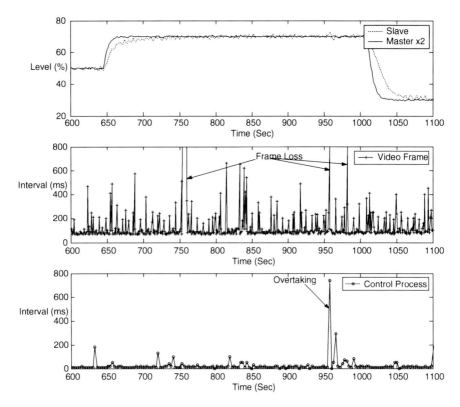

Fig. 6.6 Experimental results

control data transmission was guaranteed. There was a overtaking occurred at the point around 950 s for the control data transmission. Because of the timeout notification mechanism, this overtaking did not affect the control performance as shown in Fig. 6.6.

As the result, the network infrastructure shown in Fig. 6.2 has fulfilled the six requirements for ideal network infrastructure for Internet-based control.

6.6 Summary

Any network infrastructure for Internet-based control system ideally should meet six requirements: real-time, reasonably reliable, timeout notifying, priority enabled, time synchronization, and the ability to pass through a firewall. But the reality is far from perfect and causes Internet transmission delay and data loss. This chapter presents a network infrastructure for Internet-based control, in which TCP, UDP, RCP, and NTP are integrated together to satisfy the above six requirements. The communication process starts by using TCP to initialize a connection.

The communication process continues with RCP beginning real-time transmission by binding the underlying UDP to TCP. Afterwards, the communication relies on RCP only, and TCP only provides functions to detect the absence of members. NTP is for time synchronization. The implementation of the network infrastructure in the chapter consists of three phases: endpoint initialization phase, service initialization phase, and service processing phase. The experimental results show that the six requirements for Internet-based control have been preliminarily fulfilled.

References

Chen, X., and Yang, S.H., (2002), Control Perspective for Virtual Process Plants, *Proceedings of The 5th Asia-Pacific Conference on Control and Measurement*, China Aviation Industry Press, Beijing, China, pp. 256-261.

Comer, D.E., (2006) Internetworking with TCP/IP, Vol1, 5/E, Prentice Hall.

Han, K.H., Kim, S., Kim, Y.J., and Kim, J.H., (2001) Internet control architecture for Internet-based personal robot, *Autonomous Robots*, 10, pp. 135–147.

Luo, R.C., and Chen, T.M., (2000) Development of a multi-behavior-based mobile robot for remote supervisory control through the Internet, *IEEE Transactions on Mechatronics*, 5(4), pp. 376–385.

Magyar, B., Forhecz, Z., and Korondi, P., (2001) Observation and measurement of destabilization caused by transmission time delay through the development of a basic tele-manipulation simulation software, *IEEE WISP*, pp. 125–130.

Mills, D.L., Thyagarajan, A., and Huffman, B.C., (1997), Internet timekeeping around the globe. *Proceedings of the Precision Time and Time Interval (PTTI) Applications and Planning Meeting*, pp. 365–371.

Overstreet, J.W., and Tzes, A., (1999) An Internet-based real-time control engineering laboratory, *IEEE Control Systems Magazine*, 19, pp. 320–326.

Perkins, C., (2003) RTP-Audio and Video for the Internet, Addison-Wesley.

Rybaczyk, P., (2005) Expert Network Time Protocol: An Experience in Time with NTP, APRESS.

Chapter 7
Dealing with Internet Transmission Delay and Data Loss from the Control Perspective

7.1 Overcoming the Internet Transmission Delay

Performing remote control over the Internet is making global-range remote control systems a reality. This new type of control system allows for the remote monitoring and adjustment of industrial plants over the public Internet. With the pervasiveness of the Internet, plants stand to benefit from new ways to retrieve data and react to plant fluctuations from anywhere around the world and at any time. A pictorial schematic of an Internet-based control system is shown in Fig. 7.1. The remote operator may view what is happening via a Web camera or an image display.

The total time for performing an operation (a control action) per cycle is composed of four parts:

1. t_1 time delay for making control decision by a remote operator
2. T_f time delay for transmitting a control command from the remote operator to the local system (the Web server)
3. t_2 execution time of the local system to perform the control action
4. T_b time delay for transmitting the control information from the local system to the remote operator

Previously, in Chaps. 2–5, we have discussed a number of new features in the design of Internet-based control systems, such as requirement specification, architecture design, interface design, and real-time data transfer. In Chap. 6, we considered one of the most important problems in the case of remote control, the time delay that appears in the information transmission. Chapter 6 was concerned with overcoming the time delay and data loss from the *network* view. This chapter focuses on how to overcome the Internet transmission delay and data loss from the *control* perspective, including control system architecture design and time delay compensation. Much literature (Zhivoglyadov and Middleton 2003; Montestruque and Antsaklis 2003; Magyar et al. 2001; Yang et al. 2005a; Liu et al. 2005) exists in this area. The majority of publications (Zhivoglyadov and Middleton 2003; Montestruque and Antsaklis 2003) adopt the model-based output feedback control approach. Some of them (Yang et al. 2005a; Liu et al. 2005) focus on how to design a model-based time delay compensator or a state observer to reduce the

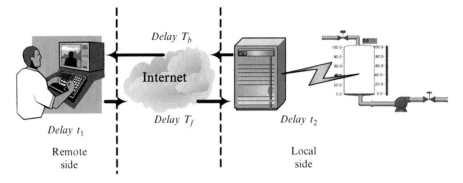

Fig. 7.1 Internet-based control system (Yang et al. 2003)

effect of the transmission time delay. These approaches maintain a process model in the remote site to approximate the plant behaviour during the time periods when the time delay or data package loss causes sensor data being not available. This process model is incorporated in the controller process and is sometimes updated regularly. Whenever time delay and data loss occur in the data transmission, the plant is controlled in open loop, and all measurements are generated by the plant model. When the network communication is recovered, the plant is controlled in closed loop. The difficulty is that the plant model must be able to accurately describe the behaviour of the plant; otherwise the open-loop control will not function correctly over a long period. It is somewhat excessive to have to model the Internet time delay for accurate prediction at every instant. This chapter aims to provide more practical solutions than the model-based approach to overcome the effect of the Internet transmission delay and data loss for Internet-based control systems.

7.2 Control Structure with the Operator Located Remotely

The control structure with the operator located remotely shown in Fig. 3.8 is one practical approach for Internet-based control, which is insensitive to time delay. Yang and Chen (2003) named this control structure as Virtual Supervision Parameter Control (VSPC). As shown in Fig. 7.2, the detailed control functions are implemented in the local control system. Internet-based control over VSPC is invoked only when updated parameters like setpoints and Proportional-Integral-Derivative (PID) parameters are required to be sent to the local control system. The new set of VSPC parameters is used as input for the local control system during the next iteration by VSPC until another set of parameters is received. One of the advantages of VSPC is the preservation of any investment in legacy systems, such as scheduling Internet-based control systems through existing DCS and PLC. Also, VSPC offers a high safety level, because the local control system will be working as normal even

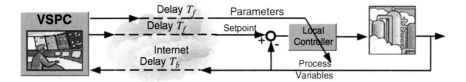

Fig. 7.2 Virtual supervision parameter control

if the Internet is disconnected from them. Furthermore, it is unlikely to be greatly affected by any Internet time delays because the Internet time delays T_f and T_b, as shown in Fig. 7.1, are excluded from the closed loop of the control system shown in Fig. 7.2. When VSPC is not able to meet the requirements specified, wider control functions should be shared by Internet-based control. For example, the remote user should be able to invoke a model-based advanced controller implemented in the local control computer. In this case, the output of the Internet control level may be a start-up command and/or several initial conditions for the model-based advanced controller.

Maintaining the existing level of safety for the process plant and the local control system at the abnormal status is essential for the VSPC to be applied in any real plant. The real challenges are how remote operators are able to detect an emergency situation and how to avoid the emergency control function being affected by Internet time delays and the abnormal status of the network systems. The VSPC runs a "better safe than sorry" safety model, which means that if anything goes wrong in the plant, the local control system, or the network systems, local operators have the authority to move the VSPC into a mode in which any command from the remote side will be ignored.

7.3 Internet-based Control with a Variable Sampling Time

As a consequence of the nature of Internet transmission, the architecture of the control system entails the network service transmitting the control signal and the functional structure accepting the control functions. In this section, a novel Internet-based control architecture is proposed to minimize the effect of the Internet time delay and data loss, which is based on the canonical one, with the controller located remotely as shown in Fig. 3.9.

It is arguable that the traditional discrete control functional structure is unsuitable for Internet-based control, which uses a fixed sampling time and the execution time is predictable as shown in Fig. 7.3a. Since the Internet transmission delay time is unpredictable, the execution time of the traditional discrete control is not predictable, which could lead to serious problems. One problem, which could arise, is that the control information arrives just after the sampling time has passed.

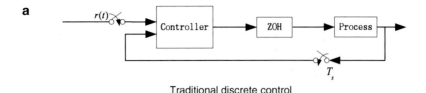

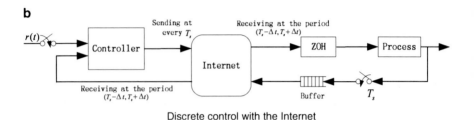

Fig. 7.3 Discrete and Internet-based control functional structures

As a result, the control system could lose a number of control signals and/or potentially increase the delay time because of the fixed sampling time. Therefore, it is necessary to develop a new functional structure to deal with the uncertain execution time.

Conceptually, the architecture should include the underlying network services and the control functional structure. Here, we only address the control functional structure based on the developed network service targeted for the control system (Chen and Yang 2002). A new Internet-based control functional structure as shown in Fig. 7.3b could evolve from the traditional discrete control structure as shown in Fig. 7.3a, which maintains the main function of this latter structure, such as Zero-Order Hold (ZOH). If the discrete control structure is considered as a tight coupling structure, the new structure as shown in Fig. 7.3b is a loose coupling structure, which introduces a tolerance time Δt to handle the unpredictable Internet communication. The value of the tolerance time is less than the sampling time, so that the control law can still be maintained. Instead of a single point receiving control signals, the new structure allows for a bounded period of time to receive and at every sampling time to transmit, which potentially maximizes the opportunity of a successful transmission. It should be noted that the sending/receiving actions are triggered by the timers synchronized by the network service.

For a long time delay and data loss, the structure employs the signal buffer at the feedback channel. In any control system, the control command becomes useless after an unexpected long time delay, which can be considered as noise. In contrast, the historical feedback signal could be still useful, particularly for correcting the model. We will use the historical feedback to compensate the control action in Sect. 7.5. In other words, the structure ignores the timeout control command signal and guarantees the delivery of the feedback signal.

7.4 Multi-rate Control

7.4.1 Two-level Hierarchy in Process Control

As shown in Fig. 3.5, any plant-wide process control system involves four levels of control. From bottom to top, the four levels are the process and protection level, the basic regulatory control level, the advanced control level or called supervisory level, and the overall plant optimization or supervision level. The work in Chap. 3 (Yang et al. 2003) introduced the Internet as an extra level and suggested that the Internet can be connected to any level according to the control requirements. If the Internet is connected with the regulatory control level and the advanced control level is located at the remote site to cooperate with other plants, the control system will have a two-level hierarchy as shown in Fig. 7.4, in which one controller is located at the local site, and another at the remote site, and linked through the Internet. Usually the controller at the local site is responsible for the regulation of normal situations. Once the performance of this controller is degraded due to disturbances from the environment or a change in the production situation, the controller in the remote site is used to tune the parameters and/or change the desired input for the controller at the local site. This two-level control structure has been widely used in robotic teleoperation (Han et al. 2001; Yang 2005; Yang et al. 2004; Luo and Chen 2000; Sayers 1999), which uses the controller in the local site to control the slave device and uses the one in the remote site to control the master device. One of the obvious advantages

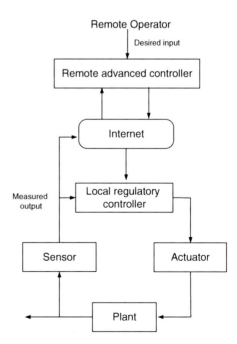

Fig. 7.4 Two-level hierarchy of process control

of using the two-level hierarchy in an Internet-based control over using the single level one, which most of the existing networked control systems use, is that should the network communication collapse for a period of time, the plant would still be under control through the local regulatory control system. Comparing Fig. 7.4 with the canonical Internet-based control structure shown in Fig. 3.10, it is obvious that Fig. 7.4 follows the canonical structure with bilateral controllers.

As described in Sect 7.1, Internet time delay is one of the major obstacles in making Internet-based process control systems a reality. It can destabilize the control system. Luo and Chen have tested the transmission efficiency of the Internet by repeatedly sending 64 bytes data from their Web server to different remote Web servers (Luo and Chen 2000). The resulting statistics of the experiments show that the Internet poses serious and uncertain time delay problems.

Considering the pictorial schematic of the Internet-based control system shown in Fig. 7.1, when the control system is in an automatic mode, the time delay for making a control decision by a remote operator, t_1, does not need to be considered. The execution time of the local system to perform the control action, t_2, can also be excluded from consideration because it was inherited from the local system and can be overcome in a traditional way. Therefore, if the feed-forward and feedback time delays T_f and T_b that appear in the information transmission over the Internet are always constants, then the Internet-based control has a constant transmission time delay. Unfortunately, as shown in Luo and Chen (2000) this is not the case. The Internet time delays, *i.e.* T_f and T_b, increase with distance and also depend on the number of nodes traversed and the Internet load (Han et al. 2001).

The network communication delay means that remote operators can no longer rely on their reflexes to detect and correct problems that happen to the controlled object. In the aeronautical and space industry such as NASA, intelligent autonomous control, which follows the principle of the two-level hierarchy shown in Fig. 7.4, has been widely employed (Sayers 1999). The intelligence of the local control system is sufficient to enable it to autonomously and immediately react to problems. The local control system is normally designed as a fast controller. The communication between remote operators and the controlled object is only carried out at a more abstract level, rather than at a detailed command level. Therefore, the remote operation can be at a lower frequency and performed via a delayed low-bandwidth communication link.

7.4.2 Multi-rate Control

Being distinguished from the existing approaches of networked control (Yang 2006), the multi-rate control scheme (Yang and Yang 2007) described in this section investigates overcoming the Internet time delay from the control system architectural point of view by enhancing the intelligence of the local control system. The multi-rate control scheme incorporates the above two-level control architecture, the lower level of which is running at a higher frequency to stabilize the plant and guarantees that the plant is under control even if the network communication is lost for a long time. The higher level of the control architecture implements the global

7.5 Time Delay Compensator Design

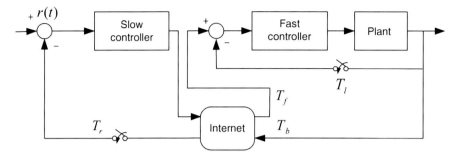

Fig. 7.5 Dual-rate control scheme

control function and is running at a lower frequency to reduce the communication load and increase the possibility of receiving data on time. We denoted the local controller as the fast controller and the remote controller as the slow controller. The structure of a dual-rate control is illustrated in Fig. 7.5. The two sampling intervals for the fast and slow controllers, T_r and T_1, are chosen as follows: $T_r = nT_1, n \in \{2, 3, \ldots\}$

The slow controller is linked with the fast controller and the plant via the Internet. The total Internet-induced transmission delay T_{delay} is equal to the sum of the transmission delays occurring in the feedback and feed-forward channels, i.e. $T_{\text{delay}} = T_f + T_b$. There are two cases involved in the above dual-rate control scheme.

Case 1: $T_{\text{delay}} + T_1 < T_r$

The time scheme of Case 1 is illustrated in Fig. 7.6. The transmission delay occurring in the local control system has been omitted, i.e. the transmission time between Nodes D and E is 0. If the sum of the total transmission delay T_{delay} and the sampling interval of the fast controller T_1 is less than the sampling interval of the slow controller T_r consequently there is no data loss during each sampling interval. Therefore, the transmission delay has no influence on the slow controller.

Case 2: $T_{\text{delay}} + T_1 \geqslant T_r$

The time scheme of Case 2 illustrated in Fig. 7.7a–g has the same meaning as in Fig. 7.6. Since the sum of the transmission delay T_{delay} and the sampling interval of the fast controller T_1 is greater than the sampling interval of the slow controller T_r the sample will be delayed and arrive at the slow controller after the next control instant. A compensator must be employed in this case to compensate for the effect caused by the transmission delay. The detail is discussed in the following section.

7.5 Time Delay Compensator Design

Any type of controllers, including PID controllers, can be implemented in the multi-rate control structure proposed in Sect. 7.4; however, a Dynamic Matrix Controller (DMC) (Cutler and Ramaker 1980) is chosen in the design of the two

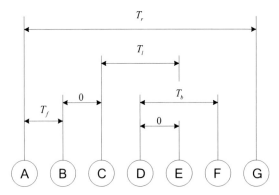

Fig. 7.6 Time scheme of dual-rate control with the total transmission delay less than the sampling interval for the slow controller (**A**) the instant at which the control action is sent out by the slow controller from the remote site; (**B**) the instant at which the fast controller receives the control command from the slow controller; (**C**) the instant at which the fast controller sent out the control action; (**D**) the instant at which the sensors send data to the controllers; (**E**) the instant at which the fast controller receives the data from the sensors; (**F**) the instant at which the slow controller receives the data from the sensors; and (**G**) the instant at which the slow controller is ready to send out the next control command

Fig. 7.7 Time scheme of dual-rate control with the transmission delay greater than the sampling interval for the slow controller

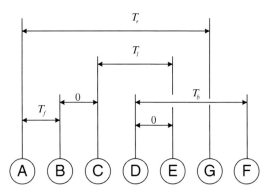

time delay compensators shown in Fig. 7.8 because the DMC has been widely accepted in the industry. The process shown in Fig. 7.8 represents a general process, which includes the local controller, and the process for a multi-rate control scheme. The compensator (Yang et al. 2005b) in the feedback channel with a data buffer is designed to overcome the time delay occurring in the transmission from the local site to a remote site. The compensator (Yang et al. 2005b) in the feed-forward channel is designed to overcome the time delay occurring in the transmission from a remote site to the local site. All the data in the Internet-based control system are sent over the Internet, provided with a time stamp generated by a global timer. The receiving time will be compared with the time stamp for each data to determine if a delay has occurred or the transmission is normal.

7.5 Time Delay Compensator Design

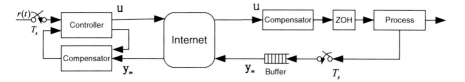

Fig. 7.8 Structure with time delay compensation

Assume that the closed-loop general process is described in the step response model as follows:

$$y(t) = \sum_{i=1}^{\infty} g_i \Delta u(t-i), \qquad (7.1)$$

where y is the process output variable, Δu is the increment of the control action, g_i is the coefficient of the step response, and t is the current time instant. The DMC general control law can be given as (Camacho and Bordons 1999) follows:

$$\mathbf{u} = (\mathbf{G}^T\mathbf{G} + \lambda \mathbf{I})^{-1} \mathbf{G}^T (\mathbf{w} - \mathbf{f}) \qquad (7.2)$$

in which λ is the penalization factor for the control costs, $\mathbf{I}$ is the unit matrix, and the superscript T means the transpose of the vector. The system dynamic matrix $\mathbf{G}$ is defined as (7.3), in which the elements are taken from the coefficients of the step response model of the process, $g_1, \ldots, g_p$ shown in (7.1):

$$\mathbf{G} = \begin{bmatrix} g_1 & 0 & \cdots & & 0 \\ g_2 & g_1 & 0 & \cdots & 0 \\ \vdots & \vdots & \ddots & & \vdots \\ g_m & g_{m-1} & \cdots & g_2 & g_1 \\ \vdots & \vdots & \ddots & \vdots & \vdots \\ g_p & g_{p-1} & \cdots & g_{p-m+2} & g_{p-m+1} \end{bmatrix}_{p \times m} \qquad (7.3)$$

where m is the control horizon and p is the prediction horizon. The control action $\mathbf{u}$ is defined as follows:

$$\mathbf{u} = [\Delta u(t) \quad \Delta u(t+1) \quad \cdots \quad \Delta u(t+m-1)]^T \qquad (7.4)$$

The reference trajectory vector $\mathbf{w}$ is defined as follows:

$$\mathbf{w} = [w(t+1) \quad w(t+2) \quad \cdots \quad w(t+p)]^T \qquad (7.5)$$

The free response vector $\mathbf{f}$ is defined as follows:

$$\mathbf{f} = [f(t+1) \quad f(t+2) \quad \cdots \quad f(t+p)]^T \qquad (7.6)$$

The free response $f(t+k)$ is computed as follows:

$$f(t+k) = y_m(t) + \sum_{i=1}^{N}(g_{k+i} - g_i)\Delta u(t-i), \qquad (7.7)$$

where $y_m(t)$ is the measurement of the process output variable and N is the process horizon. The reference trajectory $w(t+k)$ is computed as follows:

$$w(t) = y_m(t) \quad w(t+k) = \alpha w(t+k-1) + (1-\alpha)r(t+k) \\ k = 1,\ldots,N \qquad (7.8)$$

where α is a parameter between 0 and 1 for adjusting the speed of tracking and $r(t+k)$ is the setpoint of the controller.

7.5.1 Compensation at the Feedback Channel

The time delay occurring in the transmission from the local site to a remote site causes the controller at the remote site to be unable to receive the feedback signal $y_m(t)$ from the local site on time. Once the time delay occurs $y_m(t)$ in (7.7) and (7.8) will be replaced with the predictive value $\hat{y}(t|t)$, which is calculated from the step response model in (7.1) based on the available measurements at the instant t. When the transmission recovers the accumulated error $e(t)$ during the period of the time delay is used to compensate the effect. Assuming that the period of the time delay is D, consequently, the accumulated error $e(t)$ during the period of the delay is expressed as follows:

$$e(t) = \sum_{i=1}^{D}\left(y_m(t-i) - \hat{y}(t-i|t-i)\right) \qquad (7.9)$$

The compensation item chosen in this study is the accumulated error multiplied by an adjustable parameter β between 0 and 1. This compensation item is added into the free response $f(t+k)$ shown in (7.7). The history data are stored in the data buffer.

Overall, the computation of the free response and the reference trajectory can be summarized as follows:

if the time delay occurs

$$\begin{cases} f(t+k) = \hat{y}(t|t) + \sum_{i=1}^{N}(g_{k+i} - g_i)\Delta u(t-i) \\ w(t) = \hat{y}(t|t) \\ w(t+k) = \alpha w(t+k-1) + (1-\alpha)r(t+k) \\ k = 1,\ldots,N \end{cases} \qquad (7.10)$$

7.5 Time Delay Compensator Design

otherwise, if the transmission is normal

$$\begin{cases} f(t+k)=y_m(t)+\beta\sum_{i=1}^{D}\left(y_m(t-i)-\widehat{y}(t-i|t-i)\right)+\sum_{i=1}^{N}(g_{k+i}-g_i)\Delta u(t-i) \\ w(t)=y_m(t) \\ w(t+k)=\alpha w(t+k-1)+(1-\alpha)r(t+k) \\ k=1,\dots,N \end{cases}$$

(7.11)

Equations (7.10) and (7.11) form the compensator at the feedback channel.

7.5.2 Compensation at the Feed-forward Channel

The objective of the compensation at the feedback channel is to reduce the effect of the lack of any control signal as a consequence of the transmission delay. In (7.4) **u** is a vector composed of the $m-1$ future control increments. Normally, only the one at the current instant, $\Delta u(t)$, is used; the future control increments from $\Delta u(t+1)$ to $\Delta u(t+m-1)$ are simply not used. Therefore, it is possible to use these available future control increments in the situation where the time delay occurs.

When the time delay occurs in the feed-forward channel, the elements in the control vector **u** in (7.4) are shifted one step forward at every sampling interval. Equation (7.4) becomes (7.12), in which the first element is $\Delta u(t+1)$, and the last element is replaced with zero. If the time delay is longer than $m-1$, **u** will be a zero vector after delaying $m-1$ steps, and the system will be in an open control mode.

$$\mathbf{u} = \begin{bmatrix} \Delta u(t+1) & \Delta u(t+2) & \cdots & \Delta u(t+m-1) & 0 \end{bmatrix}^T \quad (7.12)$$

Using an out-of-date control signal when a feedback delay occurs is only an empirical solution to the feedback time delay. Ideally, the control action shown in (7.12) guides the process in the right direction so that the effect of the missing control signal can be reduced. However, this could lead to serious problems such as the process becoming unstable because of the mismatch between the predictive model and the actual process or a large potential process disturbance. These influences can be partly overcome through adjusting the control horizon m. In the worst-case scenario, the control horizon can be set as one, so that the control signal is maintained at a fixed value, which means that the control system is operated in an open-loop control mode. The following experimental work and simulation will demonstrate these compensation methods.

7.6 Simulation Studies

The objectives of the simulation study are to investigate the effect of the Internet time delay and data loss on the control system and to evaluate the performance of the two compensators at the feedback and feed-forward channels. The major advantage of using this simulation study includes (1) isolating the control issues from the Internet communication; (2) amplifying the frequency of the Internet time delay and data loss; and (3) providing an identical circumstance for the evaluation.

7.6.1 Simulation of Multi-rate Control Scheme

Simulation has been carried out for single-rate and dual-rate control schemes. The process is represented as a discrete transfer function $0.3/(z-1)$. The fast controller is designed as a PID controller with the parameters $K_p = 5$, $K_I = 1.2$, $K_D = 0.001$. The slow controller is designed using the control law shown in (7.4–7.8) with the parameters $N = 12, \lambda = 0.8, \alpha = 0.5, \beta = 1$. The compensators are designed according to (7.10–7.12). The sampling intervals of the two controllers T_{local} and T_{remote} are chosen as 1 and 10, respectively, for different situations.

Figure 7.9 illustrates the response to the remote setpoint change for the single-rate control scheme with a constant transmission delay $T_{delay} = 5$. The sampling interval T_s is 1 and 10 for the respective controllers. A quite large delay has been shown in the response if a lower control frequency is used. Obviously, the greater the control interval, the poorer the control performance.

Figure 7.10 shows the simulation result for the dual-rate control scheme with a larger constant transmission delay $T_{delay} = 18$ and two sampling intervals $T_{local} = 1$ and $T_{remote} = 10$. The setpoint is compared with the responses with and without compensations. If no compensation is implemented the delayed feedback is directly used as a current measure of the process output. Obvious delay in the response has been illustrated. If compensation (7.10–7.12) is used, the predictive output based on the process model is used as the current measure. Concerning the mismatch between the process model and the actual process, the available delayed feedback and the predictive output at that delayed instant are used to correct the predictive output at the current instant. The simulation results shown in Fig. 7.10

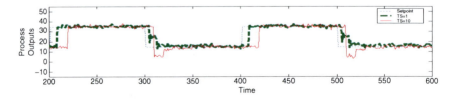

Fig. 7.9 Single-rate control with $T_s = 1$ and $T_s = 10$ (Yang and Dai 2004)

7.6 Simulation Studies

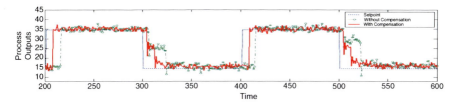

Fig. 7.10 Dual-rate control with and without feedback delay compensation (Yang and Dai 2004)

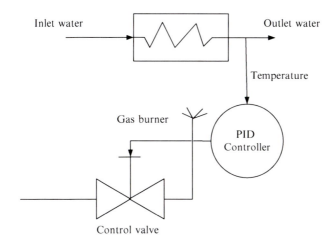

Fig. 7.11 Water heater

clearly demonstrate that the compensation has reduced the delay in the response and the dynamic performance is much better than that without the compensation.

7.6.2 Simulation of Time Delay Compensation with a Variable Sampling Time

The setpoint step change and the step disturbance have been introduced in the following simulation in order to assess the control system performance.

7.6.2.1 Simulation Design

Consider a water heater where the cold water is heated by means of a gas burner as shown in Fig. 7.11. The outlet temperature depends on the energy added to the water from the gas burner. Therefore, this temperature can be controlled by the valve, which manipulates the gas flow to the heater. Coefficient g_i can be obtained directly from the response and is the response of the outlet temperature in the water heater. It can be seen that the output stabilizes after 30 periods (model horizon $N = 30$), so the model is given by

$$y(t) = \sum_{i=1}^{30} g_i \Delta u(t - i). \qquad (7.13)$$

The response corresponds to the closed-loop controlled general process with a transfer function given by

$$G(z) = \frac{0.2713 z^{-3}}{1 - 0.8351 z^{-1}}, \qquad (7.14)$$

where the coefficients g_i are shown in Table 7.1.

The simulation study is conducted in the MATLAB®/SIMULINK® environment. The system structure is shown in Fig. 7.12, including two compensators, two delay elements to simulate the Internet, a DMC as the remote advanced controller, and the closed-loop controlled general process model. The sampling interval T_s is chosen as 1 s; the tolerant time Δt as 0.3 s. In order to provide an identical simulation circumstance for various simulation tasks, an identical time delay pattern was employed at both the feedback and feed-forward channels, which was randomly generated. Figure 7.13a illustrates the feedback delay pattern; Fig. 7.13b shows the feed-forward delay pattern. The maximum time delay is set as 10 s. In the simulation study, the prediction horizon p is chosen as 10; the control horizon m 5;

Table 7.1 Step change response

g_1 0	g_2 0	g_3 0.271	g_4 0.498	g_5 0.687	g_6 0.845	g_7 0.977	g_8 1.087	g_9 1.179	g_{10} 1.256
g_{11} 1.320	g_{12} 1.374	g_{13} 1.419	g_{14} 1.456	g_{15} 1.487	g_{16} 1.513	g_{17} 1.535	g_{18} 1.553	g_{19} 1.565	g_{20} 1.581
g_{21} 1.592	g_{22} 1.600	g_{23} 1.608	g_{24} 1.614	g_{25} 1.619	g_{26} 1.623	g_{27} 1.627	g_{28} 1.630	g_{29} 1.633	g_{30} 1.635

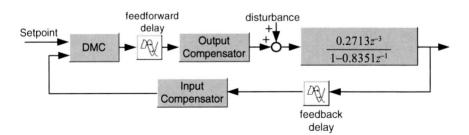

Fig. 7.12 Simulation structure (Yang et al. 2005b)

7.6 Simulation Studies

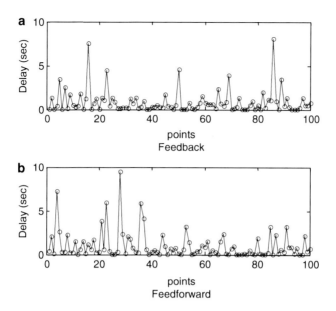

Fig. 7.13 Input and Output delay pattern

the reference trajectory parameter, $\alpha 0.7$; the parameter $\beta 1$. The setpoint step change is set as 1, and the step disturbance as 0.5.

There are four proposed scenarios: there are no time delay and data loss; the time delay and data loss exist only at the feedback channel; the time delay and data loss exist only at the feed-forward channel; and the time delay and data loss exist at both channels. Corresponding to the real world, the first scenario represents the ideal situation; the second one represents Asymmetrical Digital Subscriber Lines (ASDL) communication where the feed-forward channel has a high bandwidth; the third one represents ASDL communication where the feedback channel has a high bandwidth; and the last one represents a symmetrical communication.

7.6.2.2 Simulation Results

Figures 7.14, 7.15, and 7.16 give the simulation results for the above four scenarios. The no delay and data loss scenario is used as a reference in other three scenarios. Comparisons have been made between no time delay, time delay without compensation, and time delay with compensation in each scenario.

Figure 7.14a, b shows the process response and the control signal for the scenario, in which the time delay and data loss exist only at the feedback channel. The first disturbance is caused by the setpoint step change and the second by the step disturbance. Since the simulation circumstance is kept identical in all the simulations, it is assumed that the differences in the responses are purely dependent

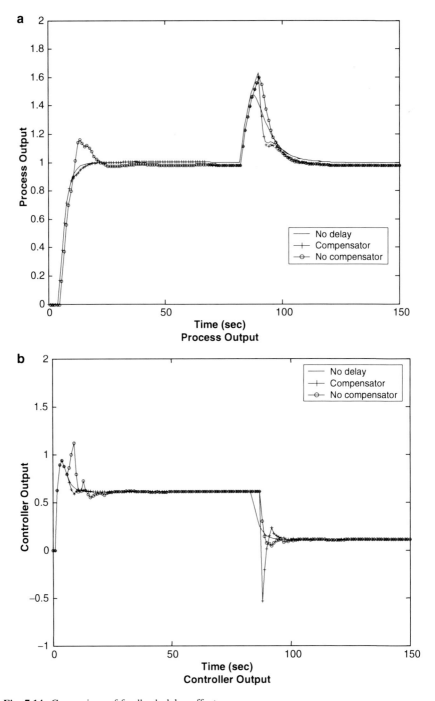

Fig. 7.14 Comparison of feedback delay effect

7.6 Simulation Studies

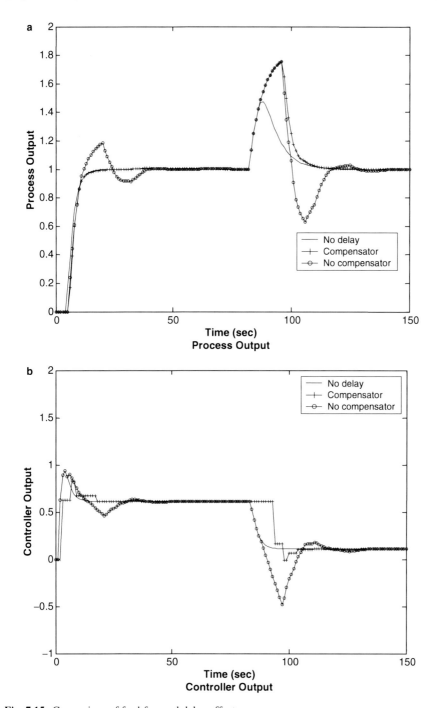

Fig. 7.15 Comparison of feed-forward delay effect

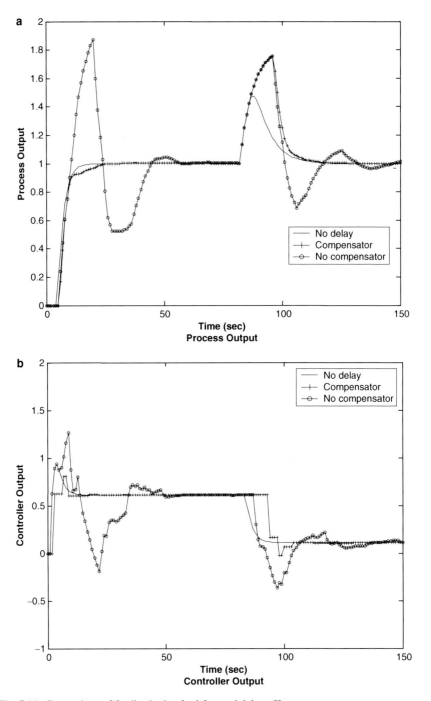

Fig. 7.16 Comparison of feedback plus feed-forward delay effects

7.7 Experimental Studies   85

upon the feedback delay and data loss and the employment of the compensator. Figure 7.14a shows that the feedback compensator can help the system to quickly follow the step change in the setpoint and significantly reduce the influence caused by the step disturbance even though there are still some difference between the one with the compensator and the one with no time delay. Figure 7.14b shows that the control signal is in a reasonable range.

Similarly, Fig. 7.15a, b gives the process response and control signal for the case, in which the time delay and data loss only exist at the feed-forward channel. The time delay and data loss cause the process variable to oscillate, which might potentially lead to the process becoming unstable. The effect of the time delay in the response with the compensator has been dramatically reduced in the setpoint step change and the step disturbance. However, the effect has not been fully compensated.

Figure 7.16a, b represents the process response and control signal for the case, in which the time delay and data loss exist at both channels. As expected, outputs become worse (more oscillation and overshoot), although they finally reach a stable value. The response with the compensators maintains the performance at an acceptable level. Comparing Figs. 7.14b, 7.15b, and 7.16b, it can be seen that more control energy is required in this case.

Comparing Figs. 7.14a and 7.15a, the effect of the time delay at the feed-forward channel is greater than the one at the feedback channel and is more difficult to be compensated. Therefore, the feed-forward channel might need a high bandwidth. It matches the current situation in most computer networks.

7.7 Experimental Studies

In this section, we implemented two Internet-based control systems in the process control laboratory at Loughborough University. The first one adopted the canonical control structure with the operator located remotely, which we named as VSPC. The second one implemented the canonical control structure with bilateral controllers. The remote controller chose a large control interval and the local one a small control interval.

7.7.1 Virtual Supervision Parameter Control

To illustrate the VSPC described in Sect. 7.2, a water tank system in the process control laboratory at Loughborough University has been chosen for the control system implementation and its evaluation. The water tank system was a teaching rig installed with a local control system. An extra Internet control level will be added into the existing local control system. With the limitation of the local control system, the link with the Internet from the plant-wide optimization level is unable to be illustrated in the test bed. The architecture of the system, including the hardware structure and the software structure, and the implementation of the system are given in this section.

7.7.1.1 Hardware Structure

As shown in Fig. 7.17, the whole system consists of five parts: a water tank, a data acquisition (DAQ) instrument, a Web server, a Web camera, and several Web clients including mobile clients, which enable wireless connection with the Internet. The tank is filled by the inlet flow controlled by a manual valve and is emptied into a drainage tank through a connection pipe and a pump. The outlet flow is controlled by a local control system located at the server to maintain the liquid level of the tank at a desired value. The DAQ instrument is in charge of A/D and D/A conversion, which converts the analogue signal of the liquid level sensor into a digital value and converts the digital value of the valve opening into an analogue value to drive the valve. The local control system of the tank is located in the server machine. The server machine and the DAQ instrument are connected and wired by RS-232c serial cables. Through the serial cable, the real-time data are exchanged between the server machine and the instrument. A Web camera connected to the server machine provides visual information to the users through a video server. Because the Web camera is independent from the DAQ, it can be considered as an extra sensor. The server provides the standard control functions as well as the Internet services and acts as the video server. The Internet service is implemented mainly based on the LABVIEW™ G-Server (NI, 2000). The main reasons for this implementation are to use Virtual Instruments (VI) provided by the LABVIEW™ to

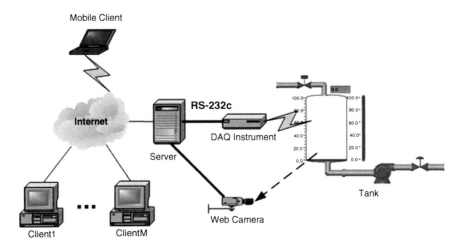

Fig. 7.17 Hardware structure of the VSPC (Yang et al. 2003)

7.7 Experimental Studies

communicate through the Internet and building a Web interface. In addition to the standard Internet service, the server also needs to establish connections between the clients and the local controller. Using a Web browser, several remotely located users are allowed to simultaneously monitor and control the tank.

7.7.1.2 Software Structure

The software of the system can be divided into two parts: the client side and the server side. Whilst the client side interacts with users, the server side is not only a Web server, but also includes the control and DAQ program to achieve the control task.

From a functional perspective, there are two programs in the client side as shown in Fig. 7.18 for the controlling and monitoring functions, which have as interfaces the control panel and the monitor panel, respectively. The control panel responds to interactions from users. The users can use it to issue commands and/or change the parameters of the controller. Using the TCP protocol, the control panel establishes the connection with the server. In addition to sending information to the server, it is also necessary to receive information from the server. If any client changes the parameters of the controller or issues a command, the server will broadcast the change to every registered user. The control panel deals with this information in order to synchronize the change and indicate the correct status of the controller. The monitor panel provides two functions: the dynamic image and the video and chat system. The dynamic image consists of graphic information including the process flowchart and the dynamic trends of process variables, which provide the essential information of the current system status. Unlike a normal web page image,

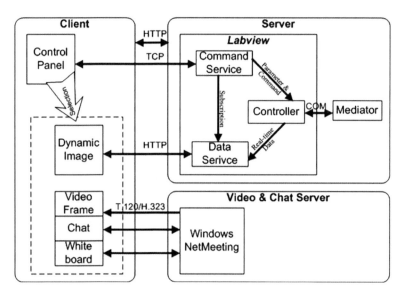

Fig. 7.18 Software structure of the VSPC (Yang et al. 2003)

the dynamic image is regularly generated by the server according to the system status, sent to the clients, and is automatically refreshed after a certain period of time. This provides clients with real-time information about the system. In order to achieve the above functions, the server push mechanism has been used. The basic principle of the server push mechanism is that the information sending action is based upon information changes, monitored by the server, rather than the client request. This not only speeds up client information updating, but also reduces the server loading. The video and chat system is designed to provide visual information for monitoring the facility and the communication channels for multi-users. Multi-users can chat to each other by sending a text message and/or sharing a white board.

On the server side, the service can be divided into two parts: the command service and the data service. The command service handles incoming requests and interprets the received information with respect to parameter values and commands for the controller and the data service. It also broadcasts the incoming information to every registered client in order to synchronize the client information. In addition, it also handles multiple client conditions such as concurrent user access. The data service is designed mainly to generate an image according to the client requirement and to send the image embedded in an html page to clients. The data that are used to generate the image are obtained from the controller. The mediator establishes a bridge between the controller and the instrument. Because the COM mechanism is fundamental for the Windows OS, and also efficient for local communication, the COM communication mechanism is chosen to support information exchange between the controller and the mediator. While COM communication is considered as a high-speed communication, the communication between the mediator and the instrument is a low-speed process. Therefore, the major task for the mediator is to coordinate the different speed components. The controller deals with standard automatic setpoint and manual control.

7.7.1.3 System Implementation

The system is implemented using a Java applet and LABVIEW™ VIs. The LABVIEW™ programs consist of three parts: the front panel, which acts as the man–machine interface; the related program, which consists of a block diagram; and the icon/connector, which is responsible for data flow between subroutines. Together, these three elements form a VI, the basic element of a LABVIEW™ program. Online process visualization over the Internet, *i.e.* dynamic image in Fig. 7.18, was realized by running G-Server software. The G-server is a hypertext-transfer-protocol (http) compatible server software, which makes hypertext markup language (html) documents and VIs located on the server available on the Internet, thus providing direct access to front panels involved in online process monitoring, *i.e.* the monitoring panel in Figs. 7.19, 7.20, and 7.21. As mentioned in the software structure described in the previous section, the server push mechanism has been used here to promote the client information updating speed and reduce the server loading. Remote control over the VSPC through the Internet was achieved by programming a Java applet using the TCP communication protocol implemented in the form of VIs.

7.7 Experimental Studies

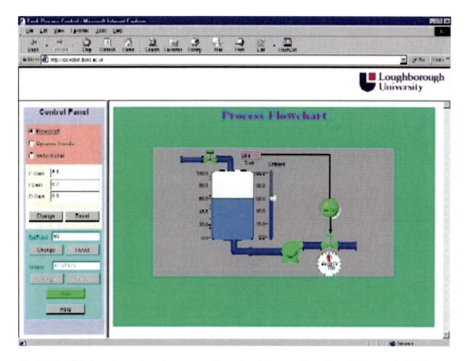

Fig. 7.19 Web-based user interface: control panel and process flowchart (Yang et al. 2003)

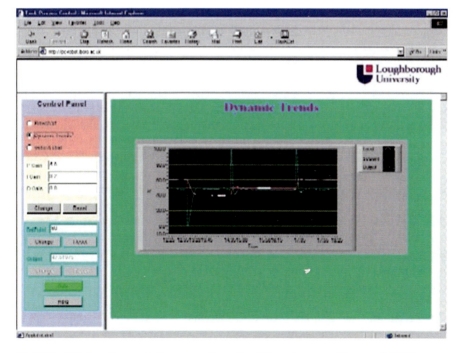

Fig. 7.20 Web-based user interface: control panel and dynamic trends (Yang et al. 2003)

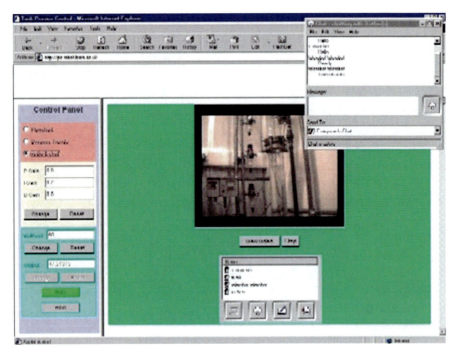

Fig. 7.21 Web-based user interface: control panel and video and chatting channel (Yang et al. 2003)

TCP enables the control panel to cope with changes made by any Web-based user to synchronize the change and indicate the correct status of the controller.

Figures 7.19, 7.20, and 7.21 illustrate the remotely located users' interface, which is divided into two parts. The left-hand side is the control panel, and the right-hand side is the monitoring panel. The control panel is a java applet where Web users can issue the command and/or change parameters of the controller to control the tank. The output of the controller is also displayed in the control panel. Because the control system can be accessed by multiple users simultaneously, the relevant latest information should be available to all the operating staff; otherwise some of them may make decisions based on out-of-date information. Therefore, all the information in the control panel will be updated immediately whenever any other registered user has made any changes to indicate the correct status of the controller. The monitoring panel is switched among the process flowchart (Fig. 7.19), the process trends (Fig. 7.20), and the video and chat panel (Fig. 7.21). The dynamic trends in Fig. 7.20 show the process responses under setpoint changes. The experimental results show that by using VSPC, the Internet-based process control system can have a behaviour similar to the local control system even with some Internet traffic delay. Figure 7.21 illustrates how the video provides the remote users with the visual information of the process. Text chat and whiteboard pop windows are invoked by pressing a corresponding button below the video, which provide users with a communication channel for cooperation.

7.7.2 Dual-rate Control with Time Delay Compensation

7.7.2.1 System Architecture

The experimental system layout is shown in Fig. 7.22. The process to be controlled through the Internet is the water tank described previously. The control target is to maintain the liquid level of the water tank at a desired value. The outlet flow is controlled by a local PID control system to maintain the liquid level of the tank at a desired value. The predictive controller is located at the remote control system, which is deployed on a laptop computer. Its function is to change the setpoint of the local PID controller. The DAQ instrument and a Web camera are the same as the one used for the VSPC in the previous section. The local control system acts as a Web server and also as a video server. The Web server provides the Internet services (IIS 5.0) and establishes connections between the remote control system and the local control system. The remote control system is connected to the Internet through a telephone modem providing 33.6 kb bandwidth. Due to the limited transmission capability of the modem, Internet congestion is often encountered. The time delay, which is the experimental circumstance that we use as input for the remote predictive controller with the time delay compensation, is also observed.

The block diagram of the above experimental system is shown in Fig. 7.23 with the feedback and feed-forward compensators. In practice, the dual-rate control scheme is more realistic and safer than the one in which a direct remote control over the Internet is exercised. As described in Chap. 3 of this book, the Internet-based process control system is intended to enhance rather than replace ordinary

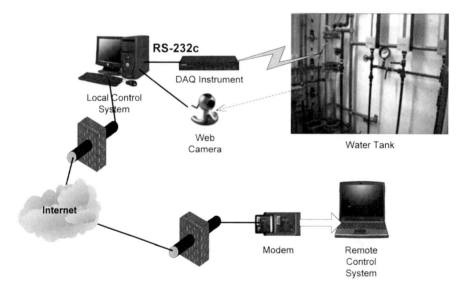

Fig. 7.22 Physical layout of the Internet-based DMC/PID dual-rate control system (Yang and Yang 2007)

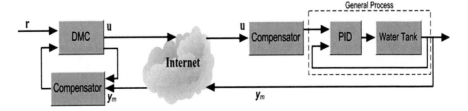

Fig. 7.23 DMC/PID dual-rate control system of the water tank (Yang and Yang 2007)

control systems by adding an extra Internet level to the control system hierarchy. The local control system ensures that, at any situation including an Internet crash, the process is still under control and safe. Another more important factor of embedding a local control system in the Internet-based control system structure is that it is very hard for the process industry to accept the idea of complete remote control over the Internet.

7.7.2.2 Experimental Results and Analysis

Three categories of the experiments have been conducted from two different locations, 5 km (in the same city) and 50 km (in the same country) away from the water tank process. The first category of experiments locates the DMC controller with the local PID controller at the local site when only local network communication is involved. The second category of experiments locates the DMC controller at a remote site when Internet communication is involved but no time delay compensation is employed. The third category of experiments is the same as the second one, but the proposed time delay compensation is applied. The experiments in the first category are used as a standard reference for comparison, in which the local network communication effect is completely ignored.

The step response model is obtained by applying a step change in the setpoint of the local PID controller, which is the model of the extended process that is the water tank plus the local PID control loop. In order to evaluate the controller performance, the setpoint (reference) of the remote DMC controller is driven by a square wave with the wave centre at 50%, which is the desired value of the liquid level of the water tank. For the control parameter, the prediction horizon p is 10, the control horizon m is 5, and the reference trajectory parameter α is 0.7. The sampling intervals of the local PID controller and the remote DMC controller are 50 and 500 ms, respectively.

The Transmission Control Protocol/Internet Protocol (TCP/IP) communication protocol is used to implement the remote communication over the Internet. The TCP/IP link between the remote DMC controller and the local PID controller has been used for the whole period of the experiments. For the symmetrical communication network, the feedback and feed-forward channels possess an equal bandwidth and have a similar latency. Only the time delay in the feedback channel during the

7.7 Experimental Studies

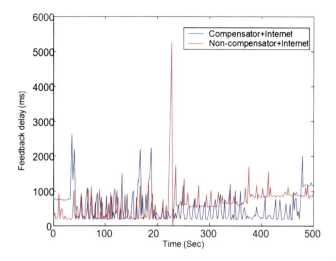

Fig. 7.24 Feedback transmission delays in experiments 2 and 3 (Yang and Yang 2007)

experiments carried out from 5 km away from the water tank is illustrated in Fig. 7.24 in order to show the uncertainty of the Internet latency. There are some but not significant differences in the time delays between the experiments carried out from 5 and 50 km distances. The reason for this might be that the amount of information actually exchanged over the Internet is small, only a few variables are communicated over the Internet, and the TCP/IP communication protocol is employed.

The experimental results are summarized in Table 7.2. Some of them are illustrated in Fig. 7.25. In order to compare the experimental results obtained under different network communication conditions and at different locations, all the experiments listed were carried out by introducing an identical setpoint change, driven by the square wave shown in Fig. 7.25 as a solid line. Three elements for each category of the experiments are recorded in Table 7.2: average feedback transmission time, standard deviation, and data loss. The average feedback transmission time indicates the data transmission time from the local site to the remote site, the standard deviation represents the dispersant degree of the transmission time, and data loss records the times of the transmission failure out of a total number of transmissions. Data loss could be caused by the transmission data loss and/or the transmission timeout. There are a number of criteria that can be used for controller performance evaluation. Only the Integral Square Error (ISE) criterion is employed in Table 7.2.

For the experiments carried out at a 5 km distance from the water tank, the total Internet time delay is about 1 s, which is double the average feedback transmission times for experiments 2 and 3, 494.32 ms and 555.60 ms. The total Internet time delay is greater than the sampling interval of the remote DMC controller (500 ms in these experiments). The compensation shown in (7.6) and (7.7) is required. The high standard deviation values, 327.17 ms and 377.92 ms illustrate the existence of the unpredictability of the Internet transmission. The feedback transmissions in

Table 7.2 Summary of the Experimental results

Experiment order number	Conditions	Experimental results			
		Control transmission			
		Average feedback transmission time (ms)	Standard deviation (ms)	Data package loss number / total package number	Control result (ISE)
Experiments carried out locally					
1	Traditional controller + local network communication	11.95	23.85	0/250	9,561.9
Experiments carried out from a 5 km distance					
2	Non-compensation controller + Internet communication	494.32	327.17	13/217	10,923
3	Compensation controller + Internet communication	555.60	377.92	20/205	9,957.7
Experiments carried out from a 50 km distance					
4	Non-compensation controller + Internet communication	553.06	357.62	16/210	11,012
5	Compensation controller + Internet communication	582.36	460.75	26/194	9,980.8

experiments 2 and 3 over the Internet have 13 out of 217 and 20 out of 205 data loss events, respectively. It can be viewed that the Internet circumstance for experiment 3 is worse than that for experiment 2. Investigating the ISE values of experiments 1–3, the ISE value increases from 9,561.9 to 10,923 because of the Internet time delay if the compensation is not employed, but from 9,561.9 to 9,957.7 if the compensation is employed in experiment 3. However, experiment 3 shows that even in a worse case compared with experiment 2, the ISE value of the experiment with the compensation, 9,957.7, is still less than that of the experiment without the compensation. Therefore, it empirically shows the efficiency of the compensation technique.

Considering experiments 4 and 5 that were carried out at a 50 km distance from the water tank, very similar results to experiments 2 and 3 were obtained. There are some minor differences among experiments 2–5 in the Internet circumstance such as the average feedback transmission time, standard deviation, and data package loss times. Also, the control performance index ISEs for experiments 4 and 5 are very close to those for experiments 2 and 3. A similar phenomenon can be observed: even though the Internet circumstance for the experiment with the compensation (experiment 5) is worse than that without the compensation (experiment 4), the control performance of experiment 5 is still slightly better than that of experiment 4.

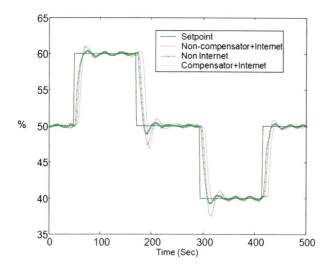

Fig. 7.25 Control performances of experiments 1–3 (Yang and Yang 2007)

Two findings can be observed by comparing the results of experiments 2 and 3 with those of experiments 4 and 5: (a) the performance of the Internet-based control system would be independent of its physical location if the Internet circumstance has not been significantly changed; and (b) the control performance with the time delay compensation is better than that without the compensation even in the worse Internet circumstances.

Figure 7.25 gives the graphical presentation of the experimental results for experiments 1–3. Experiment 1 was carried out in the local network and is used as a standard reference for comparison, in which the network communication effect is completely ignored. It is observed that the experimental results with the compensation over the Internet have less overshoot and more quickly approach the desired setpoint than the ones without the compensation.

7.8 Summary

Internet transmission delay is one of the biggest obstacles in the design of Internet-based control system. Since the number of nodes and the Internet load affect the Internet time delay, it is variable and unpredictable. Therefore, a control architecture, which is insensitive to the time delay, is needed. From the control system structure aspect, this chapter gives five control system structures, which are less sensitive to the time delay than traditional ones. For example, the VSPC control scheme excludes the Internet time delay from the closed loop of the control system and is unlikely to be greatly affected by the Internet traffic. The multi-rate control scheme employs a two-level control hierarchy: the fast controller is located at the lower level and the slow controller at the higher level. The remote controller is running at a lower frequency to reduce the influence of the data loss and the Internet transmission load, and the local controller is running at a higher frequency to stabilize the process.

This chapter investigates the potential of using the multi-rate control scheme and the model-based compensation to overcome the Internet transmission delay. If the Internet transmission delay is less than the sampling interval for the slow controller at the higher level, there will be no data loss; otherwise compensation is required. In this case, the predictive measurement of the process output with the correction based on the available delayed measurement is used as the current measurement in the compensation.

The simulation and comprehensive experimental studies focus on the VSPC and multi-rate control scheme with time delay compensation and have illustrated that the multi-rate control scheme with the time delay compensation offers a promising way to efficiently reduce the effect of Internet time delay on control performance.

Today the high-speed Ethernet, also a non-deterministic communication medium, is being adopted for process automation. Industry is beginning to implement networked control systems through this high-speed communication medium. Given the potential development of the next-generation Internet and other enhancements to the WWW infrastructure, the speed of the next-generation Internet might be sufficiently fast to be able to dramatically reduce the transmission delay and data loss. Therefore, it might be reasonable to expect that Internet latency and data loss might become less important issues in future Internet applications.

References

Camacho, E. F., and Bordons, C., (1999) *Model Predictive Control*, Springer, London.

Chen, X., and Yang, S.H., (2002) Control perspective for virtual process plants, *The 5th Asia-pacific Conference on Control and Measurement*, China, pp. 256–261.

Cutler, C. R., and Ramaker, B. C., (1980) Dynamic Matrix Control – A Computer Control Algorithm. *In Proc. Joint Automatic Control Conference*, Paper WP5-B, San Francisco.

Han, K. H., Kim, S., Kim, Y. J., and Kim, J. H., (2001) Internet control architecture for Internet-based personal robot, *Autonomous Robots*, 10:135–147.

Liu, G. P., Xia, Y., and Rees, R., (2005) Predictive control of networked systems with random delays, in *Proc. 16th IFAC World Congress*, We-M15-TO/2.

Luo, R. C., and Chen, T. M., (2000) Development of a multibehavior-based mobile robot for remote supervisory control through the Internet, *IEEE Transactions on Mechatronics*, 5: 376–385.

Magyar, B., Forhecz, Z., and Korondi, P., (2001) Observation and measurement of destabilization caused by transmission time delay through the development of a basic telemanipulation simulation software, *IEEE WISP*, pp. 125–130.

Montestruque, L. A., and Antsaklis, P. J., (2003) On the model-based control of networked systems, *Automatica*, 39: 1837–1843.

Sayers, C., (1999) *Remote Control Robotics*, Springer, Berlin, 27–40.

References

Yang, S. H., (2005) Remote control and condition monitoring, in Chapter 8 of the book *E-manufacturing: Fundamentals and Applications*, Cheng, K. (ed.), WIT Press, Southampton, pp. 195–230.

Yang, T. C., (2006) Networked control system: a brief survey, *IEE Proceedings: Control Theory and Applications*, 7: 537–545.

Yang, S. H., and Dai, C., (2004) Multi-rate Control in Internet-based Control Systems, *UK Control 2004 Proceedings*, Sahinkaya, M. N., and Edge, K. A. (eds), Bath, UK, ID-053, CD-ROM.

Yang, L., and Yang, S. H., (2007) Multi-rate control scheme for Internet-based control systems, *IEEE Transactions on SMC (Part C)*, 37(2): 185–192.

Yang, S. H., Chen X., and Alty, J. L., (2003) Design issues and implementation of Internet-based process control systems, *Control Engineering Practice*, 11: 709–720.

Yang, S. H., Zuo, X., and Yang, L., (2004) Control system for Internet-enabled arm robots, *Lecture Notes in AI*, 3029: 663–672.

Yang, Y., Wang, Y. J., and Yang, S. H., (2005) A networked control system with stochastically varying transmission delay and an uncertain process, in *Proc. 16th IFAC World Congress*, WE-M15-TO/3.

Yang, S. H., Chen, X, Tan, L., and Yang, L., (2005) Time delay and data loss compensation for Internet-based process control systems, *Transactions of the Institute of Measurement and Control*, 27(2): 103–118.

Zhivoglyadov, P. V., and Middleton, R. H., (2003) Networked control design for linear systems, *Automatica*, 39: 743–750.

Chapter 8
Design of Multi-rate SISO Internet-based Control Systems

8.1 Introduction

Many researchers have been investigating the extent to which the sampling interval selection affects the control performance in time-triggered control systems such as networked control systems (Yu et al. 2004; Lian et al. 2002). It has been found that as the sampling interval is decreased in a distributed networked control system, although the performance initially improves, it eventually deteriorates. This is a consequence of the fact that a small sampling interval results in a heavy load on the network, which would cause long time delays or data transfer failures. Nevertheless, it has not been determined how the control performance is affected by the sampling interval. In other words, the question concerning the range of values for the sampling interval, which should be specified in the system requirements to achieve a predetermined level of control performance, remains open. Furthermore, although most of the design methods proposed so far ensure system stability, they are unable to achieve certain requirements on control performance, such as desired overshoot and settling time values for step responses. Achieving the control performance requirements of networked control systems on the Internet is constrained by the load limit. The load on the Internet, represented by the sampling interval of the control system, should be kept as small as possible; in other words, the sampling interval of the remote controller should be kept as large as possible. Therefore, there is a need to determine how to meet the control performance requirements subject to load minimization on the Internet.

This chapter describes such a load minimization design method for the multi-rate Internet-based control system with dynamic performance specifications. It resolves the stability problems with a multi-rate configuration. As illustrated in Fig. 8.1, the multi-rate control system is a two-level control architecture, the lower level of which guarantees that the plant is under control, even if the network communication is lost for some time; see (Yang and Yang 2007). The higher level of the control architecture implements the global control function. The two levels run at different sampling intervals. The lower level runs at a small sampling interval (higher frequency) to stabilize the plant, while the higher level runs at a larger sampling interval (lower frequency) to reduce the communication load and increase the

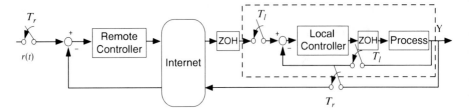

Fig. 8.1 Discrete-time multi-rate control scheme

possibility of receiving data on time. With the local system stable and the inputs of the remote controller bounded, the overall control system would remain stable. The PID controller is used for the remote control loop for simplicity and ease of tuning. The requirements on control performances, such as overshoot and settling time of step response, are represented as a pair of conjugate poles. With the dominant pole placement method, we determine the upper bounds of the remote control system's sampling interval and design the remote PID controller accordingly. The feature of this work is twofold. Firstly, it guarantees both control performance and stability of Internet-based control systems. Secondly, it simultaneously minimizes the data transmission load over the Internet, which is achieved by providing an optimal approach for maximizing the remote controller sampling interval.

8.2 Discrete-time Multi-rate Control Scheme

Consider the multi-rate Internet-based discrete-time control system shown in Fig. 8.1; it shows that the local control loop has a smaller sampling interval than the remote loop. The local controller stabilizes the plant and also meets the performance requirements of the local control system. The remote controller connected via the Internet remotely regulates the output according to the desired reference. The control input from the remote controller is forwarded to the local control system via the Internet. The feedback signal from the local control system is likewise sent to the remote controller via the Internet. We have placed two Zero-Order-Holds (ZOH) in the control system as we need to adopt two different sampling intervals there.

This transmission via the Internet inevitably brings time delay. Using the same symbols in Chap. 7, we denote the time delay of feedback via the Internet by T_b and the time delay of feed-forward by T_f. Both T_b and T_f are random variables, which are considered as the prime cause of instability and difficulty in control. However, in reality, the ranges of the time delays are known approximately. We can define them by

$$0 < T_b \leqslant \overline{T}_b, \ 0 < T_f \leqslant \overline{T}_f, \tag{8.1}$$

where, $\overline{T}_b$ and $\overline{T}_f$ are upper bounds for the Internet transmission delays of feedback and feed-forward, respectively.

8.3 Design Method

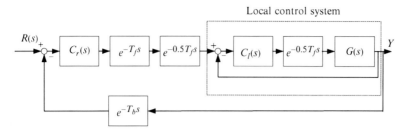

Fig. 8.2 Continuous-time multi-rate control

As defined in Chap. 7, we denote the sampling interval of the local loop by T_l and that of the remote loop by T_r in Fig. 8.1. For the sake of simplicity, we transform the above discrete-time Internet-based control system into continuous-time by approximating delay of two ZOHs as half of their sampling intervals in the controller design and presenting them as $e^{-0.5T_l s}$ and $e^{-0.5T_r s}$, respectively. The continuous-time multi-rate control scheme is redrawn in Fig. 8.2, where we replace the Internet block with two blocks of time delays: $e^{-T_b s}$ and $e^{-T_f s}$. The remote and local controllers are denoted as $C_r(s)$ and $C_l(s)$, respectively, the process as $G(s)$. The remote sampling interval is used as a measurement of load on the Internet. Load minimization for the Internet is to maximize the remote sampling interval. The overshoot and settling time values of step responses are chosen as the index of dynamic performance. Our problem, therefore, is to design the local controller and remote controller so as to minimize the load on the Internet subject to meeting the desired overshoot and settling time values of step responses.

8.3 Design Method

There are many methods to design the local controller as the type of controller to use for a given plant in the local control system is not limited and a fast sampling interval is possible. In order to reduce complexity, the local control system and the plant, *i.e.* the generic plant, are not studied directly. It is assumed that the local control system is already stable and fulfils the control specifications. The model of the local closed-loop control system can be obtained from the step response method. Thereby, we have a new and simpler block diagram as shown in Fig. 8.3, in which $G_1(s)$ is the transfer function of the the local closed-loop control system. From the dead time, overshoot, and settling time of the step responses, the local closed-loop control system is modelled as first order,

$$G_1(s) = \frac{1}{Ts+1} e^{-sL}, \qquad (8.2)$$

Fig. 8.3 Simplified block diagram of multi-rate control

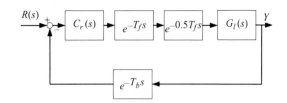

or second order,

$$G_1(s) = \frac{1}{as^2 + bs + 1} e^{-sL}, \tag{8.3}$$

with time delay L, where T, a, and b are generic process model parameters.

The remote controller can be chosen as any type of controllers. The work presented in (Huang et al. 2009) chose a dynamic matrix control (DMC) as the remote controller in which the design was based on a finite impulsive response (FIR) model. For the sake of simplicity, we choose the remote controller as a PID controller in the form of

$$C_r(s) = K_P + \frac{K_I}{s} + K_D s \tag{8.4}$$

where K_P, K_I, K_D are the PID parameters to be determined.

The PID controller in (8.4) can be rewritten in the form

$$C_r(s) = k \left(\frac{As^2 + Bs + C}{s} \right), \tag{8.5}$$

where $A = (K_D/k), B = (K_P/k), C = (K_I/k)$. The variable k is the gain of the controller and is to be determined. We use the method presented in Wang et al. (1999) for tuning of the remote PID controller and choose the controller zeros to cancel the generic process model poles, i.e. $A = a, B = b, C = 1$. Therefore, the denominator of $G_1(s)$ is cancelled with the numerator of $C_r(s)$ and the open-loop transfer function becomes

$$G_1(s)C_r(s)e^{-(T_f + 0.5T_r + T_b)s} = \frac{k}{s} e^{-(T_f + T_b + 0.5T_r + L)s} \tag{8.6}$$

For the first-order closed-loop control system, we choose a PI controller and set $K_D = 0$ in (8.4) and (8.5).

The variable k affects the performance of the closed-loop system. Using the method of dominant pole placement (Astrom and Hagglund 1995) to find a suitable k, and supposing these two poles, $p_{1,2} = -\omega_0 \zeta \pm j\omega_0 \sqrt{1 - \zeta^2}$, are dominant, i.e. to be on the root locus for the system in (8.6), where ζ is the closed-loop damping ratio, ω_0 is the undamped natural frequency, the phase condition

8.3 Design Method

$$-\omega_0\sqrt{1-\zeta^2}(T_f+T_b+0.5T_r+L)-(\pi-\cos^{-1}\zeta)=-\pi$$

has to be satisfied, giving (Wang et al. 1999)

$$\omega_0=\frac{\cos^{-1}\zeta}{\sqrt{1-\zeta^2}(T_f+T_b+L+0.5T_r)} \quad (8.7)$$

The magnitude condition then assigns the value of k in (8.6) to

$$k=\omega_0 e^{-\omega_0\zeta(T_f+T_b+L+0.5T_r)} \quad (8.8)$$

For non-time delay system, if the settling time is defined as the time for which the step response of the system reaches and stays within $\pm 2\%$ of the steady-state value, the settling time of step response can be approximately represented as (Nise 2000)

$$t_s\approx\frac{-\ln\left(0.02\sqrt{1-\zeta^2}\right)}{\zeta\omega_0} \quad (8.9)$$

The numerator of (8.9) varies from 3.91 to 4.74 as ζ varies from 0 to 0.9. An approximation for the settling time that is used for all values of ζ is as follows:

$$t_s\approx\frac{4}{\zeta\omega_0} \quad (8.10)$$

Adding in the feed-forward delay $T_f+L+0.5T_r$ in the system shown in Fig. 8.3 and replacing ω_0 with (8.7) in (8.9) and (8.10), the settling time is approximately given by

$$t_s\approx\frac{4\sqrt{1-\zeta^2}(T_f+T_b+L+0.5T_r)}{\zeta\cos^{-1}\zeta}+(T_f+L+0.5T_r) \quad (8.11)$$

When $\bar{t}_s$ is the largest allowable settling time, the range of T_r should be

$$T_r\leqslant\frac{\bar{t}_s-\frac{4\sqrt{1-\zeta^2}}{\zeta\cos^{-1}\zeta}(T_f+T_b+L)-T_f-L}{\frac{2\sqrt{1-\zeta^2}}{\zeta\cos^{-1}\zeta}+0.5} \quad (8.12)$$

Because T_b and T_f are random with certain ranges and it is impossible to find the exact value. The most conservative upper bounds of T_b and T_f denoted as $\overline{T}_b$ and $\overline{T}_f$, respectively, are applied in (8.8) to give

$$k=\omega_0 e^{-\omega_0\zeta(\overline{T}_f+\overline{T}_b+L+0.5T_r)} \quad (8.13)$$

where

$$\omega_0 = \frac{\cos^{-1}\zeta}{\sqrt{1-\zeta^2}(\overline{T_f} + \overline{T_b} + L + 0.5T_r)}$$

Equation (8.12) becomes

$$T_r \leqslant \frac{\overline{t_s} - \dfrac{4\sqrt{1-\zeta^2}}{\zeta\cos^{-1}\zeta}(\overline{T_f} + \overline{T_b} + L) - \overline{T_f} - L}{\dfrac{2\sqrt{1-\zeta^2}}{\zeta\cos^{-1}\zeta} + 0.5} \tag{8.14}$$

Experiences show that satisfactory responses are obtained if closed-loop poles of damping ratio $\zeta = 0.7071$ are chosen. By (8.14) the range of T_r becomes

$$T_r \leqslant \frac{\overline{t_s} - 5.1\overline{T_b} - 6.1\overline{T_f} - 6.1L}{3.05} \tag{8.15}$$

And substituting $\zeta = 0.7071$ into (8.13) yields

$$k = \frac{0.5}{\overline{T_f} + \overline{T_b} + L + 0.5T_r} \tag{8.16}$$

The largest allowable T_r based on (8.15) is taken to calculate k in (8.16) and design the remote PID controller as well.

8.4 Stability Analysis

In this section, we consider the stability of the overall closed-loop transfer function with respect to k given in (8.16).

Here, we use the simple stability criteria for systems with an arbitrarily time-varying, but bounded, delay, provided by Kao and Lincoln (2004). The proofs of the stability criteria can be found in Kao and Lincoln (2004) and is omitted here. Consider the single-input–single-output systems in Fig. 8.4. $G(s)$ is a plant and $C(s)$ is a controller, and the control system has an uncertain time-varying time delay. The delay can be placed anywhere in the loop, and the closed-loop system of

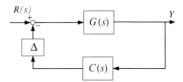

Fig. 8.4 SISO system with an uncertain time-varying time delay

$G(s)$ and $C(s)$ is stable for zero delay. The system is stable for any time-varying delays defined by $\Delta(v) = v(t - \delta(t)), 0 \leqslant \delta(t) \leqslant \delta_{max}$, if

$$\left| \frac{G(j\omega)C(j\omega)}{1 + G(j\omega)C(j\omega)} \right| < \frac{1}{\delta_{max}\omega} \quad \forall \omega \in [0, \infty] \tag{8.17}$$

The multi-rate control system shown in Fig. 8.3 gives

$$G(j\omega)C(j\omega) = \frac{k}{s} e^{-(T_f + T_b + 0.5T_r + L)s} \tag{8.18}$$

and

$$\delta_{max} = \overline{T}_f + \overline{T}_b + 0.5T_r + L \tag{8.19}$$

Substituting (8.18) and (8.19) into (8.17) yields

$$k < \sqrt{\frac{\omega^2}{\delta_{max}^2 \omega^2 - 1}} \tag{8.20}$$

To ensure the system stability under any $\omega \in [0, \infty]$, we conservatively set the stability region of k as follows:

$$0 < k < \frac{1}{\overline{T}_f + \overline{T}_b + 0.5T_r + L} \tag{8.21}$$

8.5 Simulation Studies

Consider a local system $G_1(s)$ and use the step response method to determine its transfer function. It has a step response with a dead time 2 s, overshoot 10%, and settling time 7 s. The transfer function is approximated as

$$G_1(s) = \frac{1}{0.546s^2 + 0.8737s + 1} e^{-2s} \tag{8.22}$$

Suppose the largest possible time delay caused by the Internet (round-trip time) is 1 s, which means $\overline{T}_b = \overline{T}_f = 0.5$. According to (8.5), the remote PID controller is

$$C_r(s) = k \frac{0.546s^2 + 0.8737s + 1}{s} \tag{8.23}$$

The next step is to determine k according to the largest allowable settling time. If the largest allowable settling time is 30 s, according to (8.15), the range of the sampling interval should be

$$T_r \leq 4 \qquad (8.24)$$

When the sampling interval is taken to be 2 and 6 s, respectively, and k is calculated based on (8.16), so the controller is designed as

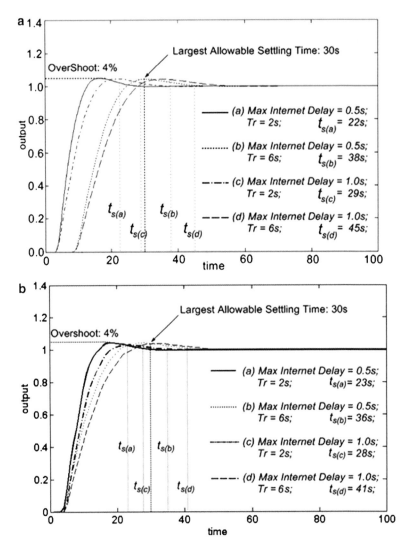

Fig. 8.5 (a) Step responses with constant delays in simulation; (b) step responses with time-varying delays in simulation (Li et al. 2010)

$$C_r(s) = \begin{cases} \dfrac{0.0683s^2 + 0.1092s + 0.125}{s}, & (k_{(a)} = 0.125) \\ \dfrac{0.0453s^2 + 0.0725s + 0.083}{s}, & (k_{(b)} = 0.083) \end{cases} \quad (8.25)$$

Likewise, when the Internet time delay (round-trip time) is 2 s and $\overline{T_b} = \overline{T_f} = 1$, according to (8.15), the range of the sampling interval should be

$$T_r \leqslant 2.23 \tag{8.26}$$

When T_r is set to be 2 and 6 s, respectively, based on (8.16), the controller should be designed as

$$C_r(s) = \begin{cases} \dfrac{0.0546s^2 + 0.0874s + 0.1}{s}, & (k_{(c)} = 0.1) \\ \dfrac{0.039s^2 + 0.0624s + 0.0714}{s}, & (k_{(d)} = 0.0714) \end{cases} \quad (8.27)$$

The step responses of these four cases with constant time delay $\overline{T_b} = \overline{T_f} = 0.5$ are shown in Fig. 8.5a. The step responses with random time delays within the range [0, 0.5] are shown in Fig. 8.5b. The obtained settling times are smaller than 30 s and the responses are satisfactory when T_r are smaller than the thresholds, i.e. $T_r = 2$. Otherwise, the obtained settling times are greater than 30 s and the responses are not satisfactory when T_r are greater than the thresholds, i.e. $T_r = 6$.

8.6 Real-time Implementation

In order to show the applicability and effectiveness of the design method described in this chapter, real-time experiments have been carried out on a real-time Process Control Unit (PCU) in the Network and Control Laboratory at Loughborough University, UK. Figure 8.6 shows the layout of the experimental system, which includes the PCU and the remote control system. Inside the PCU, there are the local control system and a water tank rig. The water tank rig consists of a process tank, sump, pump, cooler, and several drain valves. Based on the measurements of the liquid level of the water tank and flow rate of the pump, the objective is to control the liquid level or the inlet flow rate of the water tank by regulating the flow rate of the pump. The local controller parameters and sampling interval are chosen by the local operator through an operation interface. The remote control system is connected to the PCU via the Internet. More details on this experimental system can be found in Yang et al. (2007). We have conducted two sets of experiments separately, one on flow rate control and the other on liquid level control. Only the flow rate control results are given here as the liquid level control has similar results. The results of liquid level control were presented in Li et al. (2010).

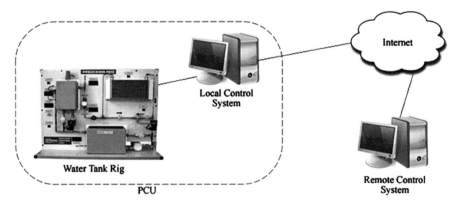

Fig. 8.6 PCU experimental system layout

The first step is to model the local control system using the step response method. A step change in the setpoint of the flow rate has been introduced into the local flow rate control system. We performed five experiments, and the step response was within 14.9% overshoot and 9.9 s settling time on average. The dead time was 0.5 s on average. Therefore, the local closed-loop control system is modeled as a second-order object with a transfer function:

$$G_1(s) = \frac{1}{2.922s^2 + 1.771s + 1} e^{-0.5s} \qquad (8.28)$$

To determine the largest possible time delay caused by the Internet, we chose a selection of IP addresses all around the globe to estimate the average value. Twenty-five proxy servers in Asian countries, including China, South Korea, Singapore, and Japan, were tested. The range of round-trip times was above 360 ms, and the maximum value was 1,059 ms. Twenty-five proxy servers in North America, including United States, Canada, and Mexico, gave an average round-trip time of 180 ms. Twenty-five proxy servers from Europe, including France, Germany, Austria, and Romania, gave an average value of 105 ms. On the basis of our measurement, the average value of Internet delay for our experiment was set to be half of the maximum value we obtained. As a result, the largest possible time delay caused by the Internet between the local and remote controllers in our experiment was considered to be 0.5 s, which means

$$\overline{T_b} = \overline{T_f} = 0.5 \qquad (8.29)$$

According to (8.5) the remote PID controller is

$$C_r(s) = k \frac{2.922s^2 + 1.771s + 1}{s} \qquad (8.30)$$

8.6 Real-time Implementation

If the largest allowable settling time was set as 15 s, when the dead time was $L = 0.5$, the range of the remote sampling interval according to (8.15) should be

$$T_r \leqslant 2.082 \tag{8.31}$$

When the remote sampling interval was taken to be 2.082 s, the largest value in order to minimize the data transmission load, and k was calculated based on (8.16)

$$k = 0.197 \tag{8.32}$$

The remote flow rate controller was designed as

$$C_r(s) = 0.348 + \frac{0.197}{s} + 0.576 \tag{8.33}$$

Implementing the remote controller in the system, we performed another five experiments. As shown in Table 8.1, the average overshoot was 6.8% and the average settling time was 7.9 s. The unit of the flow rate was litre per minute (ℓ/min). On average, the performance was satisfactory.

If the remote sampling interval was taken to be 4 s, which was out of the range of (0, 2.082), then k was calculated as

$$k = 0.143 \tag{8.34}$$

The remote flow rate controller was designed as

$$C_r(s) = 0.253 + \frac{0.143}{s} + 0.418 \tag{8.35}$$

The experiment results are shown in Table 8.2. On average, the step response of the remote controller had overshoot 14% and 15.2 s settling time. The performance was unsatisfactory as the settling time was greater than the desired value of 15 s. Figure 8.7 is the comparison of these two results observed from these two sets of experiments.

Table 8.1 Flow rate experiments ($k = 0.197$)

Exp. no.	Overshoot			Settling time		
	Setpoint (ℓ/min)	Max. value (ℓ/min)	Percentage (%)	Start point (second)	Ending point (second)	Duration (seconds)
1	1.5	1.65	10	33.5	42	8.5
2	1.5	1.68	12	78.5	82	9.4
3	1.5	1.55	3	116.6	121	6.4
4	1.5	–	–	151	158	7
5	1.5	1.64	9	189	203	8.2

Table 8.2 Flow rate experiments ($k = 0.143$)

Exp. no.	Overshoot			Settling time		
	Setpoint (ℓ/min)	Max. value (ℓ/min)	Percentage (%)	Start point (second)	Ending point (second)	Duration (seconds)
1	1.5	1.74	16	5	18.5	13.5
2	1.5	1.68	12	51	68	17
3	1.5	1.79	19.3	99	114	15
4	1.5	1.62	8	143.5	160	16.5
5	1.5	1.72	14.7	195.6	209.6	14

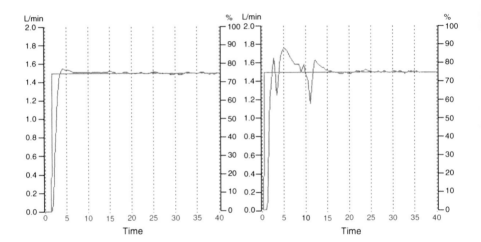

Fig. 8.7 Flow rate control comparison (Li et al. 2010)

8.7 Summary

This chapter presented a method to meet performance requirements and minimize load on the network for Internet-based control systems. The relationship between the sampling interval and settling time of the system step response was calculated for network control systems perturbed by time-varying bounded time delays. The remote PID controller was tuned to fulfil the requirement on the settling time of step response and to maximize the sampling interval, so as to reduce the network load of the control system.

As the local closed-loop control system in the multi-rate Internet-based control system is treated as a single plant, the approach avoids the complexity of large-scale system design; the required computation is not significant and is easily adopted in many networked control systems scenarios. Good responses in both simulation examples and real-time implementation have demonstrated the merits of the design approach. A necessary stability relationship for the multi-rate control system is given as an inequality condition.

References

Astrom, K.J., and Hagglund, T., (1995) PID Controllers: Theory, Design and Tuning (Second Edition), Research Triangle Park, North Caorlina, Instrument Society of America, pp. 179–193

Huang, J., Wang, Y., Yang, S.H., and Xu, Q., (2009) Robust stability condition for remote SISO DMC controller in networked control systems, *Journal of Process Control*, 19, pp. 743–750.

Kao, C.Y., and Lincoln, B., (2004) Simple Stability Criteria for Systems with Time-varying Delays, *Automatica* 40, pp. 1429–1434.

Li, Y., Yang, S.H., Zhang, Z., and Wang, Q.G., (2010) Network Load Minimization Design for Dual-rate Internet-based Control systems, *IET Control Theory and Applications*, 4(2), pp. 197–205.

Lian, F.L., Moyne, J., and Tilbury, D., (2002) Network Design Consideration for Distributed Control Systems. *IEEE Transactions on Control Systems Technology*, 10(2), pp. 297–307.

Nise, N.S., (2000) Control Systems Engineering (Third Edition), New York, John Wiley & Sons, pp. 195–196.

Wang, Q.G., Lee, T.H., Fung, H.W., Bi, Q., and Zhang, Y., (1999) PID Tuning for Improved Performance, *IEEE Transactions on Control Systems Technology*, 7(4), pp. 457–465.

Yang, L., and Yang, S.H., (2007) Multi-rate Control in Internet- Based Control Systems, *IEEE Transactions on Systems, Man and Cybernetics – Part C: Applications and Reviews*, 37(2), pp. 185–192.

Yang, S.H., Dai, C., and Knott, R.P., (2007) Remote Maintenance of Control system Performance over the Internet, *Control Engineering Practice*, 15(5), pp. 533–544.

Yu, J.Y., Yu, S.M., and Wang, H.Q., (2004) Survey on the Performance Analysis of Networked Control Systems. *2004 IEEE International Conference on Systems, Man and Cybernetics*, pp. 5068–5073.

Chapter 9
Design of Multi-rate MIMO Internet-based Control Systems

9.1 Introduction

The structure of multi-rate Internet-based control systems was given in Chap. 7. Chapter 8 presented the design of a SISO controller for this new type of control system. This chapter discusses the design of multi-rate MIMO Internet-based control systems. Figure 9.1 modifies the structure of the dual-rate control systems shown in Fig. 7.9 to show the situation where the Internet transmission delay is included in both the feedback and the feed-forward communication channels (T_b and T_f). The dual-rate control scheme has been demonstrated in a process control rig (Yang and Yang 2007; Yang and Dai 2004) and has shown a great potential to overcome Internet time delays and bring this new generation of control systems to a point where they are a viable option for industrial use. In Fig. 9.1, we show the local controller as a fast controller and the remote controller as a slow controller. The control interval of the slow controller is chosen as the control interval of the fast controller (T_s) multiplied by an integer m. If the integer m is equal to 1, the dual-rate control becomes a single-rate control. Choosing these two control intervals for the fast and slow controllers is critical. Chapter 8 discussed an approach for choosing the control interval for the slow controller to achieve both load minimization and a desired settling time. In most cases, choosing the control intervals is based on trial-and-error methods. Higher control frequency improves control performance in total but also increases communication loads on the Internet. Lower control frequency reduces communication loads but may be not able to achieve a satisfactory control performance. Furthermore, a dual-rate Internet-based control system may be unstable for certain control intervals. A sufficient and necessary stability condition for dual-rate Internet-based control systems is required to inform the system design process. For the sake of simplicity, we will only discuss a linear system where no uncertainty in the process model and time delay is considered.

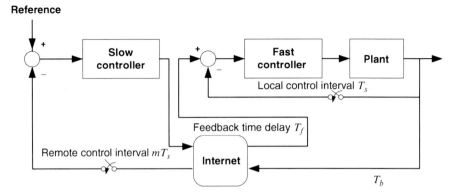

Fig. 9.1 Dual-rate control scheme with the Internet

9.2 System Modeling

In order to model the above dual-rate control scheme for a MIMO system, we choose a discrete-time state-space model to represent the plant and a state feedback control for both fast and slow controllers. The Internet is represented as two blocks of time delays: z^{-T_f} and z^{-T_b} for the feed-forward and feedback channels. respectively. Figure 9.2 shows a control block diagram for the dual-rate MIMO control scheme, where $x(k) \in R^n$ is the system state, $y(k) \in R^p$ is the system output, $u_1(k) \in R^{m_1}$ is the output of the remote controller, $u_2(k) \in R^{m_2}$ is the output of the local controller, and $r(k) \in R^{m_1}$ is the input to the remote controller. K_1, K_2 are control gain matrices with appropriate dimensions, T_f, T_b are Internet transmission delays occurring in the feed-forward and feedback channels, respectively, A_1, B_1, B_2, and C are parameter matrices of the system model where $A_1 \in R^{n \times n}, B_1 \in R^{m_2 \times m_1}, B_2 \in R^{n \times m_2}, C \in R^{p \times n}$, and k is the time index with $k \geqslant 0$ being typical.

9.2.1 State Feedback Control

For the system shown in Fig. 9.2, it is assumed that the sampling interval of the remote controller is the m multiple of the sampling interval of the local controller with m being positive integer, and the switching device SW1 closes only at the instant time $k = im, i = 0, 1, 2, \ldots$, and otherwise, it switches off. Correspondingly, remote controller $u_1(k)$ updates its state at $k = im, i = 0, 1, 2, \ldots$, only, and otherwise, it keeps invariable. Therefore, the system can be described by (9.1) with two time delays T_f and T_b.

$$\begin{cases} x(k+1) = A_1 x(k) + B_2 u_2(k) \\ u_2(k) = B_1 u_1(k - T_f) - K_2 x(k), \\ y(k) = C x(k) \end{cases} \quad (9.1)$$

9.2 System Modeling

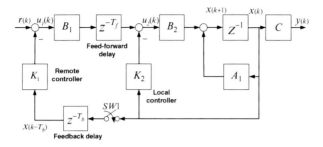

Fig. 9.2 Dual-rate state feedback control

where the output of the remote controller $u_1(k - T_f)$ is given by

$$\begin{cases} u_1(k - T_f) = r(k - T_f) - K_1 x(k - T_f - T_b), & k = im, \\ u_1(k - T_f) = r(im - T_f) - K_1 x(im - T_f - T_b), & k = \{im + 1, \ldots, im + m - 1\} \end{cases} \quad (9.2)$$

with $k = 0, 1, 2, \ldots$, $i = 0, 1, 2, \ldots$.

Substituting (9.2) into (9.1) yields that for the instant $k = im$,

$$\begin{cases} x(k+1) = (A_1 - B_2 K_2) x(k) - B_2 B_1 K_1 x(k - T_f - T_b) + B_2 B_1 r(k - T_f) \\ y(k) = Cx(k) \end{cases} \quad (9.3)$$

and for the instant $k \in \{im + 1, \ldots, im + m - 1\}$,

$$\begin{cases} x(k+1) = (A_1 - B_2 K_2) x(k) - B_2 B_1 K_1 x(im - T_f - T_b) + B_2 B_1 r(im - T_f) \\ y(k) = Cx(k) \end{cases} \quad (9.4)$$

9.2.2 Output Feedback Control

Figure 9.3 shows the output feedback for the remote controller. For the instant $k = im$, the output of the remote controller has

$$\begin{aligned} u_1(k - T_f) &= r(k - T_f) - K_1 y(k - T_f - T_b) \\ &= r(k - T_f) - K_1 C x(k - T_f - T_b) \end{aligned} \quad (9.5)$$

and for the instant $k \in \{im + 1, \ldots, im + m - 1\}$, the output of remote controller has

$$\begin{aligned} u_1(k - T_f) &= r(im - T_f) - K_1 y(im - T_f - T_b) \\ &= r(im - T_f) - K_1 C x(im - T_f - T_b) \end{aligned} \quad (9.6)$$

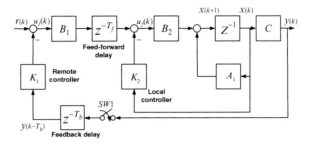

Fig. 9.3 Dual-rate output feedback control

Applying (9.5) and (9.6) into the system model described as (9.1), the corresponding closed-loop system can be given for the instant $k = im$, by

$$\begin{cases} x(k+1) = (A_1 - B_2K_2)x(k) - B_2B_1K_1Cx(k-T_f-T_b) + B_2B_1r(k-T_f) \\ y(k) = Cx(k) \end{cases} \quad (9.7)$$

and for the instant $k \in \{im+1, \ldots, im+m-1\}$,

$$\begin{cases} x(k+1) = (A_1 - B_2K_2)x(k) - B_2B_1K_1Cx(im-T_f-T_b) + B_2B_1r(im-T_f) \\ y(k) = Cx(k) \end{cases} \quad (9.8)$$

9.3 Controller Design

There are various ways to determine the two feedback control gain matrices K_1 and K_2. In this section, we employ the design of the Linear Quadratic Regulator (LQR) (Antsaklis and Michel 1997), and firstly design the local feedback control gain matrix K_2 without considering the presence of the remote controller, and then design the remote feedback control gain matrix K_1 with consideration of both feedback and feed-forward Internet transmission delays.

Consider the local control system shown in Fig. 9.2

$$\begin{cases} x(k+1) = A_1x(k) + B_2u_2(k) \\ u_2(k) = B_1u_1(k-T_f) - K_2x(k) \\ y(k) = Cx(k) \end{cases} \quad (9.9)$$

The LQR problem is to determine a control sequence $\{u_2^*(k) = -K_2x(k)\}, k \geq 0$, which minimizes the cost function

$$J(u(k)) = \sum_{k=0}^{\infty} (y^T(k)Qy(k) + u^T(k)Ru(k)) \quad (9.10)$$

9.3 Controller Design

for any initial state $x(0)$, where the weighting matrices Q and R are real symmetric and positive definite, and the superscript T denotes the transpose operator.

Assume that $(A_1, B_2, Q^{\frac{1}{2}}C)$ is reachable and observable. Then the solution to the LQR problem is given by the linear state feedback control law

$$u_2^*(k) = -K_2 x(k) = -(R + B_2^T P_c^* B_2)^{-1} B_2^T P_c^* A_1 x(k) \quad (9.11)$$

where P_c^* is the unique, symmetric, and positive-definite solution of the discrete-time algebraic Riccati equation given by

$$P_c = A_1^T (P_c - P_c B_2 (R + B_2^T P_c B_2)^{-1} B_2^T P_c) A_1 + C^T Q C \quad (9.12)$$

i.e.

$$K_2 = (R + B_2^T P_c^* B_2)^{-1} B_2^T P_c^* A_1 \quad (9.13)$$

Once K_2 is obtained, the remote control system shown in Fig. 9.2 can be represented as Fig. 9.4 and given by

$$\begin{cases} x(k+1) &= (A_1 - B_2 K_2) x(k) + B_2 B_1 u_1(k - T_f) \\ u_1(k - T_f) &= r(k - T_f) - K_1 x(k - T_f - T_b), & k = im \\ u_1(k - T_f) &= r(im - T_f) - K_1 x(im - T_f - T_b), & k = \{im+1, \ldots, im+m-1\} \\ y(k) &= C x(k) \end{cases} \quad (9.14)$$

where $k = 0, 1, 2, \ldots$, $i = 0, 1, 2, \ldots$

The system represented in (9.14) is a linear system with a delayed input $u_1(k - T_f)$, and a delayed state feedback $K_1 x(im - T_f - T_b)$. The design of the feedback control gain matrix K_1 for (9.14) can follow the procedures shown in (9.10) – (9.13) if the system model can be transferred to one with no delay in input and state feedback. One way of transformation is to extend the state from $x(k)$ to $\bar{x}(k)$, an extended state shown in (9.15).

$$\bar{x}(k) = \left[x(k)^T \ x(k+1)^T \ \ldots \ x(k+T_f)^T \ \ldots \ x(k+T_f+T_b-1)^T \ x(k+T_f+T_b)^T \right]^T \quad (9.15)$$

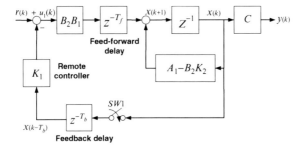

Fig. 9.4 Remote state feedback control

Gain K_1 is determined by applying (9.15) into (9.14) and then following (9.10) to (9.13).

Similarly, once K_2 is obtained, the remote control system with output feedback control shown in Fig. 9.3 can be given by

$$\begin{cases} x(k+1) = (A_1 - B_2 K_2)x(k) + B_2 B_1 u_1(k-T_f) \\ u_1(k-T_f) = r(k-T_f) - K_1 C x(k-T_f - T_b), & k = im \\ u_1(k-T_f) = r(im-T_f) - K_1 C x(im-T_f - T_b), & k = \{im+1,\ldots,im+m-1\} \\ y(k) = Cx(k) \end{cases} \quad (9.16)$$

where $k = 0, 1, 2, \ldots$, $i = 0, 1, 2, \ldots$. The output feedback control gain matrix can be obtained in a similar way. The detail is omitted here.

9.4 Stability Analysis

For the vector $x = (x_1, \ldots, x_n)^T$ and the matrix $A \in R^{n \times n}$, $\|x\|$ and $\|A\|$ denote the norms of x and A, respectively, and $\lambda(A)$ denotes the eigenvalues of A. The unit matrix is denoted as I.

For the stability analysis, we can let $r(k) = 0$, i.e. zero input to the remote controller, and then the closed-loop system with state feedback control described as (9.3) and (9.4) becomes

$$\begin{cases} x(k+1) = (A_1 - B_2 K_2)x(k) - B_2 B_1 K_1 x(k - T_f - T_b), & k = im \\ x(k+1) = (A_1 - B_2 K_2)x(k) - B_2 B_1 K_1 x(im - T_f - T_b), & k \in \{im+1,\ldots,im+m-1\} \\ y(k) = Cx(k), & k \in \{im,\ldots,im+m-1\} \end{cases} \quad (9.17)$$

Similarly, let $r(k) = 0$ and then the closed-loop system with output feedback control described as (9.7) and (9.8) becomes

$$\begin{cases} x(k+1) = (A_1 - B_2 K_2)x(k) - B_2 B_1 K_1 C x(k - T_f - T_b), & k = im \\ x(k+1) = (A_1 - B_2 K_2)x(k) - B_2 B_1 K_1 C x(im - T_f - T_b), & k \in \{im+1,\ldots,im+m-1\} \\ y(k) = Cx(k), & k \in \{im,\ldots,im+m-1\} \end{cases}$$

(9.18)

Obviously, the closed-loop systems described as (9.17) and (9.18) can be rewritten as

$$\begin{cases} x(k+1) = Ax(k) + Bx(k-\tau), & k = im \\ x(k+1) = Ax(k) + Bx(im-\tau), & k \in \{im+1,\ldots,im+m-1\}, \\ y(k) = Cx(k), & k \in \{im,\ldots,im+m-1\} \end{cases} \quad (9.19)$$

where $\tau = T_f + T_b$, $A = A_1 - B_2 K_2$, $B = -B_2 B_1 K_1$ for the system described as (9.17) with state feedback, or $A = A_1 - B_2 K_2$, $B = -B_2 B_1 K_1 C$ for the system described as (9.18) with output feedback.

9.4 Stability Analysis

In the following discussion, the initial condition of the system (9.19) is given by $x(k) = \Phi(k)$, $-\tau \leqslant k \leqslant 0$, *if round-trip delay time τ is an integer* *Otherwise,*

$$x(k) = \Phi(k), -\tau < k \leqslant 0, \quad (9.20)$$

where $\Phi(k) \in R^n$ is a vector-valued piecewise function.

Lemma (Guan et al. 2006a): The solution of (9.19) and (9.20) can be given by

$$x(im+l) = A^l x(im) + (A^{l-1} + \ldots + I)Bx(im - \tau), \, l = 1, 2, \ldots, m, \\ i = 0, 1, \ldots, \quad (9.21)$$

From (9.19) and (9.20), it is easy to obtain the following for $i = 0$.
When $k = 0$,

$$x(1) = Ax(0) + Bx(-\tau).$$

When $k = 1$,

$$\begin{aligned} x(2) &= Ax(1) + Bx(-\tau) \\ &= A[Ax(0) + Bx(-\tau)] + Bx(-\tau) \\ &= A^2 x(0) + (A + I)Bx(-\tau). \end{aligned}$$

When $k = 2$,

$$\begin{aligned} x(3) &= Ax(2) + Bx(-\tau) \\ &= A[A^2 x(0) + (A + I)Bx(-\tau)] + Bx(-\tau) \\ &= A^3 x(0) + (A^2 + A + I)Bx(-\tau). \end{aligned}$$

Similarly, when $k = m - 1$,

$$\begin{aligned} x(m) &= Ax(m) + Bx(-\tau) \\ &= A[A^{m-1}x(0) + (A^{m-2} + \cdots + A + I)Bx(-\tau)] + Bx(-\tau) \\ &= A^m x(0) + (A^{m-1} + \cdots + A + I)Bx(-\tau), \end{aligned}$$

which imply that (9.21) holds if $i = 0$ and $l = 1, 2, \ldots, m$.
For any positive integer i, when $k = im$,

$$x(im + 1) = Ax(im) + Bx(im - \tau).$$

When $k = im + 1$,

$$\begin{aligned} x(im+2) &= Ax(im+1) + Bx(im - \tau) \\ &= A[Ax(im) + Bx(im - \tau)] + Bx(im - \tau) \\ &= A^2 x(im) + (A + I)Bx(im - \tau). \end{aligned}$$

Similarly, when $k = im + m - 1$,

$$\begin{aligned} x(im + m - 1) &= Ax(im + m - 2) + Bx(im - \tau) \\ &= A[A^{m-2}x(im) + (A^{m-3} + \cdots + I)Bx(im - \tau)] + Bx(im - \tau) \\ &= A^{m-1}x(im) + (A^{m-2} + \cdots + A + I)Bx(im - \tau), \end{aligned}$$

and when $k = im + m - 1$,

$$\begin{aligned} x(im + m) &= Ax(im + m - 1) + Bx(im - \tau) \\ &= A[A^{m-1}x(im) + (A^{m-2} + \cdots + I)Bx(im - \tau)] + Bx(im - \tau) \\ &= A^m x(im) + (A^{m-1} + \cdots + A + I)Bx(im - \tau), \end{aligned}$$

which imply that the above Lemma holds for any nonnegative integer i and $l = 1, 2, \ldots, m$. This therefore completes the proof of the lemma.

Let

$$\Lambda = \left\{ \lambda \mid \det\left[\lambda^{l+\tau}I - A^l\lambda^\tau - (A^{l-1} + \cdots + I)B\right] = 0, \quad l = 1, \ldots, m \right\} \quad (9.22)$$

where det() denotes the determinant of a matrix, and λ denotes the solution of (9.22) with the matrix A, B, and the time delay τ, I denotes the unit matrix.

Theorem (Guan et al. 2006a)
The system (9.19) with the initial condition (9.20) is asymptotically stable if and only if for any $\lambda \in \Lambda$, $|\lambda| < 1$, where Λ is defined by (9.22).

Proof
For any $\lambda \in \Lambda$, let $x(k) = \lambda^k e$, where $e \in R^n$, $e \neq 0$ is an eigenvector of the matrix

$$\lambda^{l+\tau}I - A^l\lambda^\tau - (A^{l-1} + \cdots + I)B \quad l = 1, \ldots, m$$

corresponding to the eigenvalue λ. Then $x(k) = \lambda^k e$ is a solution of (9.21) if and only if

$$\lambda^{im+l}e - A^l\lambda^{im}e - (A^{l-1} + \cdots + I)B\lambda^{im-\tau}e = 0, \, l = 1, 2, \ldots, m, \, i = 0, 1, \ldots,$$

i.e.

$$\lambda^{im}[\lambda^{l+\tau}I - A^l\lambda^\tau - (A^{l-1} + \cdots + I)B]e = 0, \, l = 1, 2, \ldots, m, \, i = 0, 1, \ldots,$$

For $e \neq 0$, and $\lambda \neq 0$, the above equation can be rewritten as follows:

$$\det[\lambda^{l+\tau}I - A^l\lambda^\tau - (A^{l-1} + \ldots + I)B] = 0, \quad l = 1, \ldots, m \quad (9.23)$$

It immediately follows that the solution of (9.21) is asymptotically stable if and only if for any $\lambda \in \Lambda$, $|\lambda| < 1$, where Λ is defined by (9.22). This completes the proof of the theorem.

9.5 Design Procedure

For the multi-rate Internet-based control system (9.1) with state feedback (9.2) or with output feedback (9.6), as well as the given matrices B_1, B_2, A_1, C, time delay $\tau = T_f + T_b > 0$, and the remote sampling interval multiple $m > 0$, the problem is that how to design the feedback control gain matrices K_1 and K_2 such that the corresponding controlled system (9.19) with the initial condition (9.20) is asymptotically stable. According to the LQR control law (9.13), (9.2), and the corresponding theorem, the design procedure is given as the following two steps.

Step 1: For the system (9.1) with the given matrices B_1, B_2, A_1, C, time delay $\tau = T_f + T_b > 0$, and the remote sampling interval multiple $m > 0$, design the control gain matrix K_2 based on (9.13) and the corresponding equation for K_1, compute $A = A_1 - B_2 K_2$, $B = -B_2 B_1 K_1$ for the system described by (9.17) with state feedback, or $A = A_1 - B_2 K_2$, $B = -B_2 B_1 K_1 C$ for the system described by (9.18) with output feedback. Then go to next step.

Step 2: For $l = 1, 2, \ldots, m$, calculate the eigenvalues $\lambda \in \Lambda$ where Λ is defined by (9.22). If all $\lambda \in \Lambda$, satisfy $|\lambda| < 1$, for $l = 1, 2, \ldots, m$, then the design is completed; otherwise, go back to Step 1 and redesign the two-state feedback control gains K_1 and K_2. The traditional design methods can be used here (Antsaklis and Michel 1997).

9.6 Model-based Time Delay Compensation

This section discusses the compensation for the transmission delay in the multi-rate Internet-based control system with a state feedback, which possesses a randomly varying transmission delay.

We assume the round-trip transmission delay $\tau = T_f + T_b$ is bounded and stochastically varying, i.e. $0 \leq \tau \leq \tau_{max}$, where τ_{max} is an integer, which can be determined according to the maximum time delay under the normal network condition. In case the network temporarily collapses, the time delay will be greater than τ_{max}, and the latest available control action will keep being used until the network recovers.

On the basis of the above assumptions, the local controller may receive zero, one, or more (up to τ_{max}) control action packets from the remote controller during a single remote control interval mT_s. If the local controller receives no control action packets during any remote control interval $[t_k, t_{k+m})$, $u_1(k)$ in the last received control action packet will continue to act on the local controller during the next sampling interval $[t_{k+m}, t_{k+2m})$. If the local controller receives more than one control action packets during any sampling interval $[t_k, t_{k+m})$, only the most recent received control action packet is kept and the local controller will discard the others.

Concerning the random round-trip transmission time delay and letting the external reference $r(k) = 0$, the state feedback controller without time delay compensation as shown in (9.2) can be re-stated as follows:

$$u_1(k - T_f) = \begin{cases} -K_1 \sum_{j=0}^{\tau_{max}} \delta(\tau - j)x(k - \tau), & k = im, \\ -K_1 \sum_{j=0}^{\tau_{max}} \delta(\tau - j)x(im - \tau), & k = \{im + 1, \ldots, im + m - 1\} \end{cases} \quad (9.24)$$

$$\delta(\tau - j) = \begin{cases} 0 & \tau \neq j \\ 1 & \tau = j \end{cases} \quad (9.25)$$

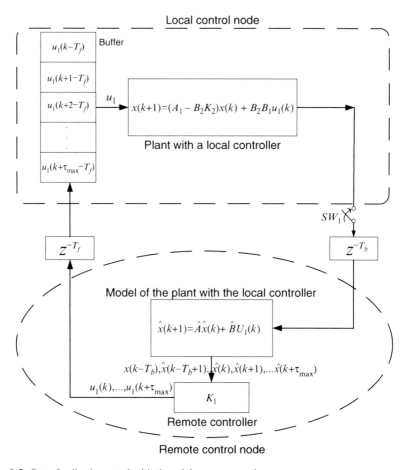

Fig. 9.5 State feedback control with time delay compensation

$$\sum_{j=0}^{\tau_{max}} \delta(\tau - j) = 1 \qquad (9.26)$$

where $k = 0, 1, 2, \ldots$, $i = 0, 1, 2, \ldots$. τ represents the random round-trip transmission time delay rounded to an integer, $\tau \in \{0, 1, \ldots, \tau_{max}\}$.

The compensation is implemented by using a buffer in the local control node and a state estimator in the remote control node. Different from the compensation method introduced in Chap. 7, this section considers a MIMO system with state feedback and randomly varying transmission delay in both feedback and feedforward channels.

For the sake of simplicity, a two-step design strategy is used here. Firstly, a state feedback controller is designed without considering the effect of the time delay, and then a remote control node and a local control node are designed to compensate the effect of the feedback delay and the feed-forward delay. Figure 9.5 illustrates the principle for the compensation of the transmission delay for the multi-rate Internet-based control. A process model is located in the remote control node in order to predict the future performance of the plant based on the latest available measured states. A buffer is located in the local control node in order to compensate for the effect of the transmission time delay.

9.6.1 Compensation of the Transmission Delay at the Feedback Channel

Suppose the latest plant state received by the remote control node is $x(k)$ and let the external reference $r(k) = 0$. The model with the local controller, which is located at the remote control node, will predict the next τ_{max} plant states based on this measured plant state $x(k)$: $\hat{x}(k+1)$, $\hat{x}(k+2)$, to $\hat{x}(k+\tau_{max})$, calculate the current and future τ_{max} control actions: $u_1(k), u_1(k+1), \ldots, u_1(k+\tau_{max})$ and then transmit them to the local control node together with the time stamp received from the local control node. The current control action $u_1(k)$, the prediction of the plant states and the future control actions at instants $k+1, k+2, \ldots, k+\tau_{max}$ based on the measured plant state $x(k)$ are as follows:

$$\begin{cases} u_1(k) = \begin{cases} -K_1 x(k), & k = im \\ -K_1 x(im), & k = \{im+1, \ldots, im+m-1\} \end{cases} \\ \hat{x}_{k+l|k} = \begin{cases} (\hat{A}^l + (A^{l-1} + \ldots + I)\hat{B}K_1)x(k), & k = im \\ (\hat{A}^l + (A^{l-1} + \ldots + I)\hat{B}K_1)x(im), & k = \{im+1, \ldots, im+m-1\}, l = 1, 2, \ldots, m \end{cases} \\ u_1(k+j|k) = -K_1 \hat{x}(k+j|k) \\ j \in [1, 2, \ldots, \tau_{max}], \; i = 0, 1, \ldots, \end{cases} \qquad (9.27)$$

where $\hat{x}(k+l|k)$ and $u_1(k+j|k)$ denote the prediction of the plant state and the future control action at instants $k+l$ and $k+i$, respectively, based on the measured state $x(k)$.

In general, if the latest available plant state received by the remote control node is $x(k-j), j \in [0,1,\ldots,\tau_{max}]$, the prediction of the plant states and the future control actions based on them at instants $k+i, i \in [1,2,\ldots,\tau_{max}]$ can be represented as follows:

$$\begin{cases} u_1(k-j) = \begin{cases} -K_1 x(k-j), & k=im, \ j=0 \\ -K_1 x(im-j), & k=\{im+1,\ldots,im+m-1\}, \ j=0 \end{cases} \\ \hat{x}_{k+l|k-j} = \begin{cases} (\hat{A}^l + (A^{l-1}+\cdots+I)\hat{B}K_1)x(k-j), & k=im \\ (\hat{A}^l + (A^{l-1}+\cdots+I)\hat{B}K_1)x(im-j), & k=\{im+1,\ldots,im+m-1\}, l=1,2,\ldots,m \end{cases} \\ u_1(k+j|k-j) = -K_1 \hat{x}(k+j|k-j) \\ j \in [1,2,\ldots,\tau_{max}], i \in [1,2,\ldots,\tau_{max}] \end{cases}$$

(9.28)

where $\hat{x}(k+l|k-j)$ and $u_1(k+j|k-j)$ denote the prediction of the plant state and the future control action at instants $k+l$ and $k+j$, respectively, based on the measured state $x(k-j)$.

9.6.2 Compensation of the Transmission Delay in the Feed-forward Channel

Once a new measured plant state $x(k-j), j \in [0, 1, \ldots, \tau_{max}]$ is received by the remote control node, the feedback time delay T_b will be calculated by comparing the current time with the time stamp received from the local control node and then rounded to an integer multiple of the sampling interval by adding a waiting time at the remote control node into the delay, i.e. $j = T_b$. On the basis of latest available plant state $x(k - T_b)$, the control actions $u_1(k), u_1(k+1), \ldots, u_1(k+\tau_{max})$ are calculated according to (9.28) and then sent to the local control node. Once the control action packet is received by the local control node, the feed-forward time delay T_f will be calculated by comparing the current time with the time stamp received from the remote control node and then rounded to an integer multiple of the sampling interval by adding a waiting time at the local control node into the delay. The control actions available for the local controller are $u_1(k-T_f), u_1(k+1-T_f), \ldots, u_1(k+\tau_{max}-T_f)$, which are saved in the buffer at the local control node. The local control node will choose $u_1(k)$ from the above list as the current control action acting on the local controller. At instant $k+1$, i.e. the next sampling instant, if there is no any updated control action packet received from the remote control node, $u_1(k+1)$ found from the control action packet

$u_1(k - T_f), u_1(k + 1 - T_f), \ldots, u_1(k + \tau_{max} - T_f)$ will be used for the plant. If more than one control action packets are received, only the packet with the latest time stamp will be saved in the buffer. This is used to deal with the situations of package disorder and package loss.

9.6.3 Unified Form of the State Feedback Control of the Remote Controller

As this case is similar to the uncompensated control action shown in (9.24), the compensated control action obtained from the remote control node and saved in the local control node at instant k can be formalized as follows:

$$u_1(k) = \sum_{j=0}^{\tau_{max}} \delta(\tau - j) u_1(k|k - \tau)$$

$$= -K_1 \sum_{j=0}^{\tau_{max}} \delta(\tau - j) \hat{x}(k|k - \tau) \quad (9.29)$$

where $k = 0, 1, \ldots,$ $\delta(\tau - j) = \begin{cases} 0 & \tau \neq j \\ 1 & \tau = j \end{cases}$, and $\sum_{j=0}^{\tau_{max}} \delta(\tau - j) = 1$.

If the network temporarily collapses, τ may be greater than τ_{max}. The latest available $u_1(k + \tau_{max} - T_f)$ from the buffer will keep being used as the control action on the local controller until the network recovers and a new control action packet is received. Equation (9.29) involved the compensation of the total transmission delay τ since it compensates for the feed-forward transmission delay T_f in the local control node and the feedback transmission delay T_b in the remote control node when predicting the current state.

9.7 Simulation Study

Consider a discrete plant (Yang et al. 2008) shown in Fig. 9.5 with the parameters below: the remote control interval is 0.5 s, which is five times of the interval of the local controller.

$$\hat{A} = \begin{bmatrix} 1.0013 & 0.05 \\ 0.05 & 1.0013 \end{bmatrix}, \hat{B} = \begin{bmatrix} 0.0013 \\ 0.05 \end{bmatrix}$$

$$C = [0.5, 0.2]$$

The total transmission delay τ is bounded and stochastically varying within $0 \leq \tau \leq \tau_{max}$, $\tau_{max} = 3$. The state feedback controller $u_1(k) = -K_1 x(k)$, where

$K_1 = [\,10.3 \quad 3.38\,]$, was first designed in advance without considering the presence of the network in this simulation study for the sake of simplicity. This standard state feedback controller is then implemented in the Internet-based control structure shown in Fig. 9.5, where time delay compensation and dual-rate control scheme are added on. The responses of the states x_1, x_2, and the output y under the square wave setpoint change are shown in Figs. 9.6–9.8, respectively. The system was initially at a steady state, i.e $x_1(0) = 0;\quad x_2(0) = 0;\quad y(0) = 0$. The setpoint shown in Fig. 9.8 is changed from 0 to 1.0 at instant $k = 0$ and then back from 1.0 to 0 at instant $k = 100$. In Figs. 9.6 and 9.7, the solid and dash lines represent the responses of the two-state variables without and with the transmission delay compensation, respectively. It is obvious that the responses with the transmission delay compensation are quicker in approaching the new steady states and

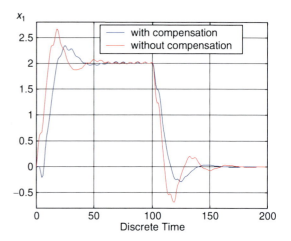

Fig. 9.6 State x_1 response

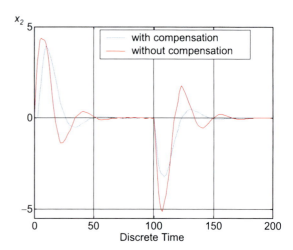

Fig. 9.7 State x_2 response

Fig. 9.8 Output response

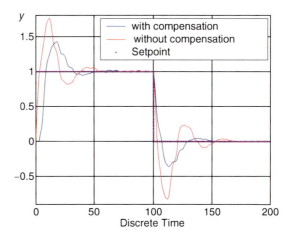

have much less overshoot. Figure 9.8 illustrating the output response achieves the same conclusion. The square wave setpoint is shown in Fig. 9.8 as a reference. The output with the transmission delay compensation has much less overshoot and approaches the setpoint more quickly than the one without the compensation. The comparison illustrates that the transmission delay compensation method introduced in Sect. 9.6 can improve the system performance.

9.8 Summary

This chapter presented the design of multi-rate MIMO Internet-based control system. Different from the previous chapter, a state-space model is used to design state feedback control and output feedback control for the multi-rate control system. The control gain matrix can be designed, in principle, by many methods, and the LQR is only introduced as an example design for the local state feedback control. A necessary and sufficient criterion of asymptotic stability for such systems is established.

The transmission delays at both feedback and feed-forward channels are compensated by introducing a buffer at the local control node to save the future control actions received from the remote control node and introducing a state-space model in the remote control node to predict the plant state that is delayed by the feedback transmission delay or is not available at the time. The remote control action actually applied to the local controller is chosen from the buffer in terms of the actual total round-trip transmission delay. The current time stamp is compared with the time stamp when the state was sent to the remote control node to compute the current total round-trip transmission delay.

This chapter only addresses the most basic case that is of linear systems with no uncertainty in their parameters. There are a number of problems, which have not be addressed in this chapter, but for which solutions can be found in the literature. Three recent research publications (Yang et al. 2008; Guan et al. 2006b, 2008) extended the results given in this chapter to the cases for multi-rate Internet-based control systems with uncertainties in the process model and multiple transmission delays. Robust passive control for Internet-based switching systems with time delay has been addressed in Guan et al. (2008). The stability of multi-rate Internet-based control systems with uncertainties and multiple time-varying delays has been addressed in Guan et al. (2006b). The stability criteria for linear systems with uncertainties or nonlinear systems are much more complicated than the theorem given in Sect. 9.4. They are discussed in Yang et al. (2008) where it is shown that they produce a Linear Matrix Inequality-based (LMI) as sufficient condition for the stability of the networked control systems under the proposed compensation scheme. Guan et al. (2006b) also gave an LMI sufficient condition for the Internet-based control systems with multiple time delay.

Some typical results from other researchers are concerned with the stability of networked control systems under various assumptions. For example, the work in Zhang et al. (2001) gave a stability criterion under the assumption that transmission delay is less than a sampling interval and the data are transmitted in a single packet. The work in Walsh et al. (2002) used Maximal Allowable Transfer Interval (MATI) between transfers of data from sensors to a controller to guarantee the networked control system stability. A necessary and sufficient condition for stability of networked control systems with state feedback and output feedback was given in Montestruque and Antsaklis (2003). Design and stability criteria of networked control systems were presented in Liu et al. (2007) in terms of model predictive control (MPC)-based networked control systems. Yang (2006) gave a review of the design of general networked control systems.

References

Antsaklis, P.J. and Michel, A.N., (1997) Linear Systems, McGraw-Hill, New York, pp. 348–349.

Guan, Z.H., Yang, S.H., and Zhang, H., (2006) Stability of Internet-based control systems with uncertainties and time-varying delay, *International Conference Control 2006 (UK Control'06)*, Glasgow, paper 215.

Guan, Z.H., Zhang, H., Yang, S.H., and Wang, H.O., (2006) Stability of Internet-based control systems with uncertainties and multiple time-varying delays, *Proceedings of the 45th IEEE Conference on Decision & Control*, pp. 6419–6424.

Guan, Z.H., Zhang, H., and Yang, S.H., (2008) Robust passive control for Internet-based switching systems with time-delay, *Chaos, Solitons, and Fractals*, 36(2), 479–486.

Liu, G.P., Xia, Y., Reeds, D., and Hu, W., (2007) Design and stability criteria of networked predictive control systems with random network delay in the feedback channel, *IEEE Transactions on Systems, Man and Cybernetics – Part C: Applications and Reviews*, 37(2), pp. 173–184.

Montestruque, L.A. and Antsaklis, P.J., (2003) On the model-based control of networked systems, *Automatica*, 39, pp. 1837–1843.

References

Walsh, G.C., Ye, H., and Bushnell, L.G., (2002) Stability analysis of networked control systems, *IEEE Transaction on Control System Technology*, 10(3), pp. 438–446.

Yang T.C., (2006) Networked control system: a brief survey, *IEE Proceedings: Control Theory and Applications*, 153(4), pp. 403–412.

Yang, S. and Dai, C., (2004) Multi-rate control in Internet-based control systems, *UK Control 2004 Proceedings*, Sahinkaya, M.N. and Edge, K.A. (eds), *UKControl'04*, Bath, UK, ID-053, [CD-ROM].

Yang, L. and Yang, S.H., (2007) Multi-rate control in Internet-based control systems, *IEEE Transactions on Systems, Man and Cybernetics – Part C: Applications and Reviews*, 37(2), pp. 185– 192.

Yang, Y., Wang, Y., and Yang, S.H., (2008) Design of a networked control system with random transmission delay and uncertain process parameters, *International Journal of Systems Science*, 39(11), pp. 1065–1074.

Zhang, W., Branicky, M.S., and Phillips, S.M., (2001) Stability of networked control systems, *IEEE Control Systems Magazine*, 1, pp. 84–99.

Chapter 10
Safety and Security Checking

10.1 Introduction

The difference between Internet-based control systems and ordinary remote control systems, tele-operation systems, and distributed control systems is that Internet-based control systems use the Internet rather than any private media as the communication medium. The Internet has both safety and security risk as a consequence of its open environment nature. The use of an Internet-based control system means that we can no longer ensure the safety on the system by just considering the local plant but have to consider the whole Internet community, as there is always a possibility that the local control system is falsified by outsiders through the Internet. Thus, aspects of the public Internet must be considered in the design of any Internet-based control systems to protect them from attack by outside hackers. Existing technologies such as plant firewall, user authentication, communication path encryption, access log, and format conversion (Furuya et al. 2000; Shindo et al. 2000) might be able to make the Internet-based control systems reasonably safe for normal use but would never be able to prevent attacks by malicious hackers. The nature of remote control also increases the safety risk to the plants since there might be no local operators at the plants. Therefore, systematic safety and security checking in the design of an Internet-based control system are essential to establish sufficient confidence in the safety levels and feel secure in using the system. Safety and security checking also aims to reduce any loss caused by the attack and provide guidelines for the operators to efficiently respond to any attacks.

By identifying the similarity of both safety and security, this chapter systematically considers the safety and security issues through the design phase and clarifies all the scenarios of malicious attacks. Actions to respond to the potential attacks are suggested as the result of the safety and security analysis.

10.2 Similarity of Safety and Security

Safety risk analysis has the aim of specifying the safety requirements of the system. Similarly, security risk analysis identifies any potential security risks. There are some differences but more similarities between the properties of safety and of security (Eames and Moffett 1999). For example, from the security perspective any weakness in the system and any dangers are called vulnerabilities and threats. In the safety analysis, weakness and dangers are called failure mechanisms and hazards, but they can be considered to be the same. In the security domains, the countermeasures that need to be put in place to counter any risks are access controls, firewalls, etc. and in safety, they are redundancy, protective equipments, monitoring devices, etc. Rushby (1994) presented the nature of safety and security, in which the differences between the two were recognized, but also illustrated how both groups subscribe to similar development techniques, *i.e.* safety and security techniques could be applicable to each other's domains. For example, security system developers could benefit from the fault tolerant approaches typically found in safety techniques and might benefit from a greater understanding of the hazard analysis methods used by safety engineers.

In general, safety, security, and their associated risk analysis techniques are closely related. Both deal with risks and result in constraints, which may be regarded as negative requirements. Both involve protective measures, and both produce requirements that are considered to be of the greatest importance. These similarities indicate that some of the techniques applicable to one field could also be applicable to the other.

In Internet-based control systems, the safety problems, because of the nature of the remote operation, are caused by the authorized users. Avoiding the failures caused by the authorized users can be achieved through safety analysis at the Internet level. The security problems are caused by the malicious attacks. Preventing attackers from accessing Internet-based control systems is ensured by the measures of network security, physical security, and data security such as firewall and data encryption. Adequate response actions should be taken to prevent any fatal accidents from happening should these measures of security fail.

10.3 Framework of Security Checking

This section presents a framework for security checking. The traditional "What-If" approach for safety checking has been used for security checking due to the similarity of security and safety checking.

10.3.1 Framework of Stopping Possible Malicious Attack

The Internet gateway is obviously the first target of any attack if a malicious hacker tries to get unauthorized access to an Internet-based control system (Shindo et al.

10.3 Framework of Security Checking

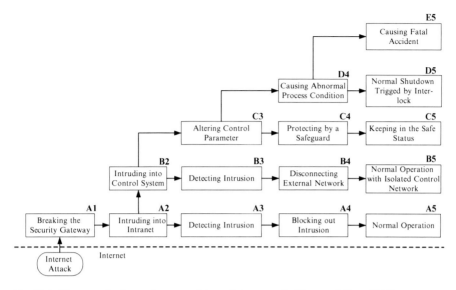

Fig. 10.1 Framework of stopping a possible malicious attack (Yang and Yang 2007)

2000). Figure 10.1 (Yang and Yang 2007) shows a comprehensive intruding path from breaking the Firewall (node A1) to causing a fatal accident (node E5) through intruding into the Intranet (node A2), intruding into the local control system (node B2), altering control parameters (node C3), and causing abnormal process conditions (node D4). Time goes from left to right and the degree of risk increases from bottom to top. Cutting off the path that starts at the node A1 and ends at the node E5, at any point, will prevent the fatal accident from happening. Figure 10.1 gives four possible points at which the path from A1 to E5 might be cut off:

- Cutting the path between the nodes A2 and B2 by detecting and shutting out the intrusion from the Intranet (nodes A3 and A4). This solution has minimum risk to the process and purely relies on the measures of the available network security and physical security. There is a rich literature (Hamdi and Boudriga 2005; Marin 2005) of solutions in these areas.
- Cutting the path between the nodes B2 and C3 by detecting the intrusion into the local control system (node B3), cutting off the link with the external network (node B4), and allowing the control system to run in isolation from the network.
- Cutting the path between the nodes C3 and D4 by using a safeguard to protect the process from an unexpected change in control parameters. The safeguard might be based on a simple threshold for a key process parameter or a complex control performance index. We discuss these safeguards in the following section.
- Cutting the path between the nodes D4 and E5 by activating a Safety Interlock Systems (SIS) to trigger the normal shutdown procedure. This is the last possible point of preventing a possible fatal accident and causes the maximal loss to the

process. The SIS has been widely used and independently implemented with safety-critical control systems (Yang et al. 2001a).

10.3.2 Framework-based What-If Security Checking

The What-If approach is basically a communication exercise and asks what-if questions about the systems or processes. Information is presented, discussed, analysed, and recorded. Specifically the potential risks are identified and determined if appropriate design measures have been taken into account to prevent an accident from happening.

The "What-If" approach was mainly used for safety checking. Because of the similarity of safety and security, the "What-If" approach can be employed for security checking according to the framework shown in Fig. 10.1. Considering the three most common security breach scenarios. Table 10.1 summarizes "What-If" security checking reviews in terms of the framework of Figure 10.1. Three actions (Actions 1, 2, and 3) are proposed and must be taken to avoid the consequences described in the "What" column occurring. In Scenario 1, the firewall and password control have been broken by malicious attacks. The corresponding action (Action 1) is to disconnect the external link between the local control system and the Internet. In Scenario 2, if a malicious attacker changed the control parameters, the control system will not work properly. The corresponding action is to trigger a safeguard system to reduce the influence of the parameter change. A safeguard system might be designed to simply filter out the abnormal control action. Scenario 3 is that, if an attacker has created an abnormal process condition, a safety interlock system (SIS) will be activated to keep the process in a safer condition and wait for the intervention from an operator. The details of three actions are described as follows:

Action 1: disconnect the external link between the local control system and the Internet.

On the one hand, an Internet-enabled plant will never be absolutely safe and secure if a remote user is allowed to directly access and make changes to the local control system. On the other hand, the plant will never be remotely controllable if a remote user is not

Table 10.1 What-If security reviews

If	What	Actions
Scenario 1: firewall and password control are broken	Attackers obtain the access to the control system	Action 1: disconnect the external link between the local control system and the Internet if the intrusion is detected
Scenario 2: attackers have modified control parameters	Disturbances have been introduced into the process	Action 2: a safeguard system filters out any abnormal change to the local control system
Scenario 3: attackers have created safety-critical conditions	A fatal accident could occur	Action 3: an emergency safety interlock system is required to be automatically activated

10.3 Framework of Security Checking

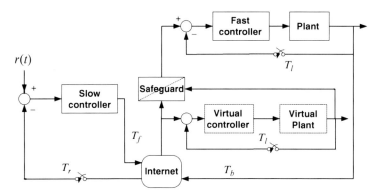

Fig. 10.2 Safe control through a virtual system

allowed to access and justify the control system. Action 1 can be implemented through a virtual controller and a virtual plant, which are introduced to act as an intermediate between the remote user and the plant. Using the dual-rate Internet-based control system described in Chaps. 7–9 as an example, Fig. 10.2 illustrates the principle of this safe control mechanism described above. Any authorized remote user can have direct access to the virtual system, but not to the real plant. The virtual system is designed to predict the behaviour of the real plant under the control action assigned by the remote user and therefore it can be used to check whether or not the control action from the remote user is doing any harm to the real plant. If the behaviour of the virtual plant under a certain remote control action is desirable, this control action is applied to the real plant via a safeguard; otherwise the remote control action is rejected.

Action 2: a safeguard system filters out any abnormal change to the local control system.

Various control performance assessment methods can be used as a safeguard to check whether or not the control action from the remote user is doing any harm to the real plant. Being simpler than an ordinary control performance assessment, the virtual process output under the control action assigned by the remote controller is used as a safeguard to calculate the *Integral of Absolute Error* (*IAE*) and *Mean Squared Error* (*MSE*).

- *Integral of Absolute Error* is the integral over time of absolute deviation between the setpoint and the measurement. It can be employed to assess the control performance during setpoint change. For a process with disturbances, this value increases monotonically with time.

$$IAE = \sum_{k=k_0}^{k_f} |e_k| \qquad (10.1)$$

where e_k is the difference between the samples of the setpoint and the output at the instant k. k_0 and k_f are the starting and ending instants of the samples.

- *Mean squared error* is the integral of the squared difference between the setpoint and the measurement divided by the time interval.

$$I_{MSE} = \frac{1}{f} \sum_{k=k_0}^{k_f} e_k^2 \qquad (10.2)$$

where $f = k_f - k_0 + 1$ denotes the number of the calculated data.

If $IAE < \lambda_1$ or $I_{MSE} < \lambda_2$, then the control action from the remote controller is acceptable, where λ_1 and λ_2 are the maximum tolerant indices, which are set according to experience; otherwise the control action received from the remote controller is a suspect.

Action 3: a safety interlock system (SIS) is activated.

A SIS is sometimes called an emergency shutdown system (AIChE/CCPS 1993) and is one of the most important protective measurements in process plants, which provides automatic actions to correct an abnormal plant event, which has not been controlled by either a basic control system or manual intervention. A SIS is only needed on those rare occasions when normal process controls are inadequate to keep the process within acceptable bounds. On the other hand, a SIS must be available to operate whenever needed. Thus, SISs must be designed independently of normal control systems and serve as a last backup system. Thorough and systematic verification of SISs is essential for safety-critical systems such as Internet-based control systems to establish the necessary degree of confidence that the total system behaves in an acceptable manner under a wide variety of process fluctuations and instrument failure conditions (Yang et al. 2001a, b).

10.4 Control Command Transmission Security

10.4.1 Hybrid Algorithm

Control command transmission security is the means of protecting control commands from change by malicious hackers during their transmission from the remote site to the local site over the Internet. Control command transmission security for Internet-based control systems must be secure enough and also satisfy the real-time requirements. In this section, we introduce a two-stage hybrid data encryption algorithm for this new type of control systems:

1. Set-up stage. In the set-up stage, the RSA (Rivest–Shamir–Adleman) algorithm (Management 2005) is used to establish the communication link by generating a RSA public key and a RSA private key and securely transferring an AES (Advanced Encryption Standard) (Daemen and Rijmen 1999) cipher. The RSA algorithm uses a public key to encrypt the AES cipher and uses a private key to decrypt it.
2. Data exchange stage. In this stage, the AES (Daemen and Rijmen 1999) is used to regularly encrypt/decrypt the real-time control commands according to the transferred AES cipher. All of the steps in the AES data encryption/decryption

10.4 Control Command Transmission Security

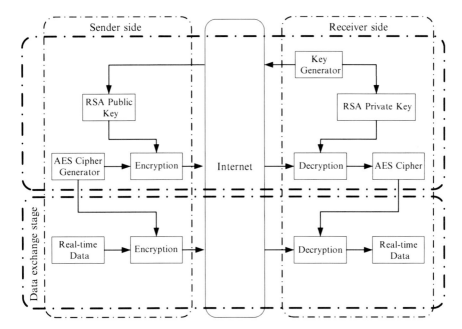

Fig. 10.3 Hybrid algorithm based on the AES and RSA algorithms

are simple matrix operations, which make the AES suitable for real-time data encryption/decryption.

The principle of this hybrid algorithm based on the AES and RSA is shown in Fig. 10.3. In the set-up stage, the RSA public and private keys are generated by the key generator on the receiver side, and then the RSA public key is sent to the sender side. The generated AES cipher on the sender side is encrypted using the received RSA public key and then sent to the receiver side and decrypted using the RSA private key. In the data exchange stage, the real-time data are encrypted/decrypted using the AES algorithm and the transferred AES cipher.

In the process of the RSA encryption, the private key is saved at the receiver site and the public key is transmitted at the same time to the sender. The RSA algorithm adopts the public key to encrypt the AES cipher and the private key to decrypt it. It is impossible to determine the private key from the public key. As the private key will never be transported, the security level of the AES cipher transferred by the RSA algorithm is much higher than that of the AES algorithm alone. The security of the RSA algorithm comes from the computational difficulty of factoring large numbers and takes much more time for encryption/decryption data than the AES one. In contrast, the AES algorithm encrypts real-time data only using simple matrix operations and has a high rate of encryption/decryption. The hybrid real-time data encryption/decryption algorithm shown in Fig. 10.3 might provide the advantages of both AES and RSA algorithms while avoiding their disadvantages.

10.4.2 Experimental Study

In order to evaluate the above hybrid encryption algorithm, the time delays produced by the AES, RSA, and their hybrid encryption/decryption algorithms are compared in a similar network environment. The total time delay is composed of three parts: the data encryption time, the encrypted data network transmission time, and the data decryption time. If the network environment and the processing power are similar, the total time delay will depend on the time spent on the data encryption/decryption. Figure 10.4 shows that the total time delay of the hybrid algorithm is close to that of the AES and is much shorter than that of the RSA. Furthermore, since the AES cipher of the hybrid algorithm is encrypted by the RSA, it is much

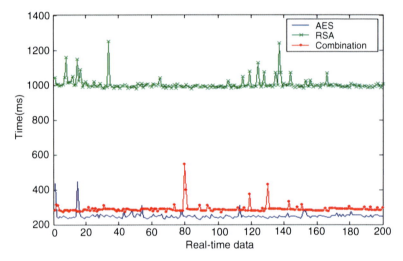

Fig. 10.4 Comparison of the end-to-end latency of AES and RSA

Table 10.2 Comparison of the experimental results

Algorithms	End-to-end real-time transmission latency			Pros	Cons
	Average latency (ms)	Maximum latency (ms)	Minimum latency (ms)		
RSA	1,003.46	1,250	978	Cipher more safer	High latency
AES	249.57	446	227	Low latency	Cipher not secure enough
Combination of AES and RSA	288.94	544	270	Cipher more secure; low latency; AES cipher and cipher-text synchronous	No authentication

hard to be deciphered. Therefore, the hybrid algorithm is securer than the AES. The experimental results are also summarized in Table 10.2. The average end-to-end latency indicates the normal operation period. The maximum and minimum latencies illustrate the existence of the unpredictability of the Internet transmission. The advantages and disadvantages of the three possible data encryption/decryption algorithms are also summarized. As shown in Table 10.2, the end-to-end latency of the hybrid algorithms is close to that of the AES algorithm and much shorter than the RSA algorithm, and the lower average latency of the hybrid algorithms, 288.94, indicates that it is suitable for securing control commands transferred over the Internet for Internet-based control systems.

10.5 Safety Checking

Due to the nature of remote operation and the Internet environment constraints, even authorized remote users may cause failures in the process without any improper operation. Therefore, it is necessary to identify what can go wrong and consider what consequence may result and prevent the potential hazards from occurring. An efficient, systematic way of finding potential risks is to introduce possible deviations from the intended design. A hazard analysis framework for computer-controlled plants was proposed in (Yang et al. 2001b; Chung et al. 1999), which was based on a Process Control Event Diagram (PCED). An extended PCED for Internet-based control systems has been introduced in Chap. 2 as a

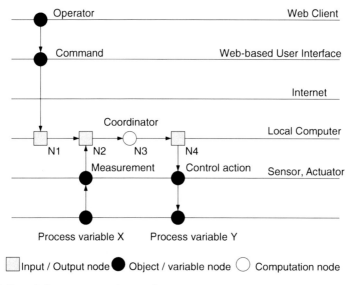

Fig. 10.5 Extended process control event diagram

functional model and shows the communication between the process, the controller, the Internet, and operators. The extended PCED, shown in Fig. 2.4, is adopted here for the safety risk analysis of Internet-based controlled plants. An example of a modified PCED is shown in Fig. 10.5. The PCED illustrates the interaction between nodes, which are arranged on six different layers (from the top to the bottom layer: Web client, Web-based user interface, Internet, Local Computer, Sensor/Actuator, and Process). As described in Chap. 2, the nodes represent the components involved in the system (e.g. sensors, actuators, and control algorithms), and an edge between two nodes represents the propagation of a signal.

The conventional HAZOP (AIChE/CCPS 1993) principle is followed here and various deviations, *i.e.* scenarios, are applied node by node in the PCED on the basis

Table 10.3 Attributes, guidewords, and interpretations for Internet-based control systems

Attribute	Guideword	Interpretation
Data/control flow	No	No information flow
	More	More data is passed than expected
	Part of	Information passed is incomplete
	Reverse	Information flow is in a wrong direction
	Other than	Information is complete, but incorrect
	Early	Information flow before it was intended
	Late	Information flow after it was required
Data rate	More	Data rate is too high
	Less	Data rate is too low
Data value	More	Data value is too high
	Less	Data value is too low
Event	No	Event does not happen
	As well as	Another event takes places as well
	Other than	An unexpected event occurred instead
Action	No	No action takes place
	As well as	Additional actions take place
	Part of	Incomplete action is performed
	Other than	Incorrect action takes place
Timing of event or action	No	Event/action never takes place
	Early	Event/action takes place before expected
	Late	Event/action takes place after expected
	Before	Happens before another expected event
	After	Happens after another expected event
Repetition time	No	Output is not updated
	More	Time between outputs is longer than required
	Less	Time between outputs is less than required
	Other than	Time between outputs is variable
Response time	No	Never happens
	More	Time is longer than expected
	Less	Time is shorter than expected
	Other than	Time is variable

of the use of "guidewords", which are words or phrases expressing specific types of deviation to systematically identify the potential safety risks.

The common guidewords in the conventional HAZOP are *no, more, less, part of,* and *other than.* The UK Ministry of Defense safety checking guidance (1996) added three more guidewords for control systems: *reverse, early,* and *late.* We add further three guidewords for Internet-based control systems: *before, after,* and *as well as.* The possible attributes of control systems are *data/control flow, data rate, data value, event, action, timing of event or action, repetition time, and response time.* The guidewords that are applicable to Internet-based control systems are shown and interpreted in Table 10.3. Deviations from normal behaviour of the control systems are considered for each attribute at each node in the PCED. Causes, corresponding consequences, and correcting actions can be proposed by a team of experts. Potential hazards or safety-critical events can then be identified. The procedure of applying the safety checking will be illustrated in the following case study.

10.6 Case Study

To illustrate how well the safety and security checking procedures can be applied, we consider the Internet-based water tank control system described in Chap. 8 as a case study, in which the control objective is to maintain the liquid level of the water tank at a desired value, the inlet flow is controlled by a local Proportional Integral Derivative (PID) controller to maintain the liquid level of the tank at a desired value, and the remote controller is designed to adjust the setpoint of the local controller from a remote site. We redraw Fig. 8.6 as Fig. 10.6.

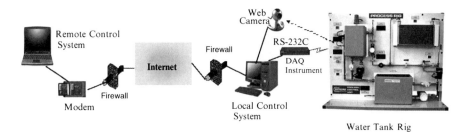

Fig. 10.6 Layout of an Internet-based control system

10.6.1 Ensuring Security

Guided by the framework of preventing possible malicious attacks presented above in Fig. 10.1, the following measures have been taken in the design of the above Internet-based control system:

- Firewall and access control. As seen in Fig. 10.6, a firewall is designed to stop any unauthorized access to the local control system. Password control is also used in the remote control system. Only registered users can view the remote control interface and have access to the Internet-based control system.
- A virtual system is employed in the local control system. The virtual system is composed of a PID control algorithm and a plant model. Any control commands received from the remote control system are firstly applied to the virtual system. The control performance index *IAE* shown in (10.1) is calculated in the virtual system to justify whether or not the control commands are potentially harming the plant. If no harm is occurring, the control commands are passed to the safeguard; otherwise, they are dropped in the virtual system.
- A safeguard is installed in the local control system. Ninety-nine percent is set as a simple threshold for the setpoint received from the remote site. Any valid setpoint must be less than or equal to this threshold; otherwise the received setpoint is omitted.
- An emergency shutdown system is running independently with the local control system. Ninety-nine percent is set as the threshold of the liquid level. Once the liquid level reaches the threshold, the inlet pump is switched off immediately; hence, the overflow of the water tank will never take place.
- All the control commands and parameters from the remote control system are encrypted before they are transferred over the Internet and decrypted after received by the local control system using the hybrid algorithm of the AES and RSA. Figure 10.4 shown in Sect. 10.4 shows the experimental results of the transmission delay of the remote setpoint for the water tank.

10.6.2 Safety Checking

If replacing the process variable X with the liquid level, the process variable Y with the opening of the inlet valve, and the computation node N3 with the PID control algorithm, the PCED shown in Fig. 10.5 then exactly describes the control logic for the Internet-based controlled water tank. Following the principles of HAZOP, deviations from a normal behaviour can be introduced node by node in the PCED by using the guidewords in Table 10.3. For example, the deviation for the action "receiving a signal from a remote site" at Node N1 in Fig. 10.5 would be "fail to receive a signal from a remote site". The consequence of this deviation is that the setpoint of the local PID controller is not available or a BAD value. If no measure

10.6 Case Study

Table 10.4 Partial HAZOP analysis results for the water tank control system

HAZOP item	Attribute	Guide word	Deviation	Causes	Consequences	Recommendations
N1	Data flow	No	No signal from the remote site	Internet congestion, Internet time delay, and Internet connection broken	Setpoint is the BAD value. The local PID controller will not work properly	The previous setpoint value is adopted if the current setpoint is the BAD value
N2	Data flow	No	No signal from the liquid level sensor	The liquid level sensor is out of order	The liquid level signal in the BAD value. The local controller will not work properly	Install a duplicate sensor
N4	Data flow	No	No signal to the outlet valve	The communication between the local computer and the outlet valve is broken	The outlet valve is left uncontrollable	Regularly checking the RS232 cable
N1	Data value	More Other than	Setpoint incorrect	The remote operator made an error or the remote system failed	The signal to the outlet valve is changed according to this incorrect setpoint	Adding a safety locking system to stop the mistake from the remote site propagating into the local control system

was taken for this consequence, the local PID controller will be not able to work properly, which may lead to the liquid level of the water tank changing dramatically. Similarly, deviations for other actions in the PCED need to be considered. Table 10.4 summarizes deviations, corresponding causes, and consequences for the water tank. Recommendations shown in Table 10.4 must be taken in order to prevent these consequences from occurring.

10.7 Summary

This chapter explores a framework of safety and security checking for Internet-based control systems. The security risk checking focuses on finding a way to prevent external malicious attacks and to prevent loss as early as possible. The safety checking aims to identify the potential hazards in Internet-based control systems. Four possible protection layers to stop malicious attacks are identified in the proposed general framework. The first layer is the standard firewall protection, which uses password control to allow only authorized users to enter the control system. The second to fourth layers are the protection measures of responding to the possible intrusion by employing a virtual system, safeguard, and SIS in order to stop any fatal accidents happening. After identifying the similarity of safety and security, the "What-If" method, which was mainly used for the safety risk analysis, is also applied to the security risk analysis. Three actions of halting any malicious attacks at different layers are proposed in terms of the What-If method. The transfer of control commands over the Internet is secured by a hybrid data encryption/ decryption algorithm, which is a combination of the ASE and RSA. An extended PCED is used to model Internet-based control systems and to derive the must-be-taken actions by applying the conventional HAZOP principle in order to ensure the system safety.

References

AIChE/CCPS, (1993) Guidelines for safe automation of chemical processes. New York: *American Institute of Chemical Engineers, Centre for Chemical Process Safety.*
Chung, P.W.H, Yang, S.H., and Edwards, D.W., (1999) Hazard identification in batch and continuous computer-controlled plants, *Industrial and Engineering Chemistry Research*, 38, pp. 4359–4371.
Daemen, J. and Rijmen, V., (1999) AES Proposal: Rijndael, available at http://csrc.nist.gov/CryptoToolkit/aes
Eames, D.P. and Moffett, J., (1999) The integration of safety and security requirements, *Lecture Notes in Computer Science*, 1698, pp. 468–480.
Furuya, M., Kato, H., and Sekozawa, T., (2000) Secure web-based monitoring and control system, *The 26th Annual Conference of The IEEE Industrial Electronics Society*, Nagoya, Japan, 2, pp. 2443–2448.

References

Hamdi, M. and Boudriga, N., (2005) Computer and network security risk management: theory, challenges, and countermeasures, *International Journal of Communication Systems*, 18(8), pp. 763–793.

DI Management, (2005) RSA Algorithm, available at http://www.di-mgt.com.au/rsa_alg.html

Marin, G.A., (2005) Network security basics, *IEEE Security and Privacy*, 3(6), pp. 68–72.

Ministry of Defense (MOD), (1996) Hazop studies on systems containing programmable electronics, Part 2: general Application Guidance, *Interim Defense Standard*, Glasgow.

Rushby, J., (1994) Critical properties: survey and taxonomy, *Reliability Engineering and System Safety*, 43, pp. 182–219.

Shindo, A., Yamazaki, H., Toki, A., Maeshima, R., Koshijima, I., and Umeda, T., (2000) An approach to potential risk analysis of networked chemical plants, *Computers and Chemical Engineering*, 24, pp. 721–727.

Yang, L. and Yang, S.H., (2007) A framework of security and safety checking for internet-based control systems, *International Journal of Information and Computer Security*, 1(1/2), pp. 185–200.

Yang, S.H., Tan, L.S., and He, C.H., (2001) Automatic verification of safety interlocks systems for industrial processes, *Journal of Loss Prevention in the Process Industries*, 14, pp. 379–386.

Yang, S.H., Stursberg, O., Chung, P.W.H., and Kowalewski, S., (2001) Automatic safety analysis of computer-controlled plants, *Computers and Chemical Engineering*, 25, pp. 913–922.

Chapter 11
Remote Control Performance Monitoring and Maintenance over the Internet

11.1 Introduction

Over the last two decades many thousand Advanced Process Control (APC) systems have been installed in process plants. The challenge facing industry now is to maintain the optimum performance of existing APC systems. The motivation for establishing remote maintenance systems is to provide efficient support for these control systems, which are characterized as being distributed either within the process plants or, on some occasions, geographically. Remote maintenance systems enable companies, with multiple sites in remote locations, to access, analyse, and react to information from the plant floor more quickly and efficiently than previously. They also enable Small and Medium Enterprises (SME) to delegate such maintenance to a service company or software supplier without the need for any internal experts. It can also virtually eliminate the need for any control software supplier's expert to conduct "on-site maintenance". Therefore, both time and money can be saved.

Internet-based control systems share various components with the remote performance monitoring systems. The relevant applications include but not limited to the following:

- Wireless and wired communications technologies, which have been used in remote health monitoring and the control of aircraft (Thompson 2004)
- The ScadaOnWeb system (2002), which targets a new standard and a generic architecture for handling numeric data on the Web and enabling process control, monitoring, and optimization via the Web
- The design of Internet-based control system (Yang et al. 2003a, 2005), which focuses on dealing with architecture selection, Internet latency, multiple user access, and security and safety
- Remote access to industrial plants, which is based on distributed object-oriented technologies (Calvo et al. 2006)
- E-Diagnostics (Sematech 2002), which enables an authorized equipment supplier's field service person to access and diagnose equipments remotely
- The health and safety issues (HSE 1995), associated with the use of the Internet and Internet-related technologies in and by industry

- Performance monitoring for control systems (Huang and Shah 1999; Nougues et al. 2002), which provides means to benchmark the performance of any system against the "best in class" performance and enables users to diagnose possible problems so that the appropriate corrective actions can be taken. A comprehensive overview of control performance assessment technology and industrial applications can be found in Jelali (2006).

There are still a number of challenges in the area. For example, what is the best means to maintain several Internet-enabled control system software components from a remote centre location? Users still appear to know little about how to identify any control system software components which is in need of repair or updating, how to maintain control system software with minimum disturbance to the controlled processes, or how to share the maintenance experience and reduce the costs between companies. They continue to see control system supplier's field service engineers wasting time and money on their worldwide travel in order to rectify a simple fault that has occurred in their real-time control software.

11.2 Performance Monitoring

Performance monitoring is a typical condition monitoring which is defined as (Institution of Production Engineers 1990): "the continuous or periodic measurement and interpretation of an item to determine the need for maintenance ... Condition monitoring is normally carried out with the item in operation, in an operable state or removed but not subject to major strip down". Condition monitoring can also be defined as (Institution of Production Engineers 1990): "the assessment of the current condition of plant and equipment by the use of techniques which can range from sophisticated computer driven instrumentation to human sensing, in order to predict failure and to economically perform maintenance only when a potential failure is identified and at a time convenient to the production schedule".

These definitions contain a number of important points. Firstly, a measurement system of operating condition is required, which should be based on an understanding of the failure mechanisms associated with the item being monitored. Secondly, an interpretation of the obtained measurement should be given to indicate the need for maintenance. The notions of prediction and maintenance introduce the most effective use of performance monitoring in practice, namely predictive maintenance. This predictive maintenance role represents the primary application of performance monitoring. If the failure is severe, it will normally be noticeable to operators by looking at the alarm page or trend plots and finding what went wrong. Therefore, the real power of the performance monitoring is to alert operators to problems before they become really noticeable.

Performance monitoring is concerned with measurement and interpretation of data to provide some informative output regarding the current performance, and predicted future performance, of a system, and therefore to maintain the system

11.2 Performance Monitoring

performance at a desirable level. Therefore, performance monitoring should include the following objectives:

- To collect the minimum amount of information needed to plan predictive maintenance and to avoid high cost breakdown maintenance
- To detect system defects and the development of faults during system operations
- To maintain the performance of the system at a desirable level

In order to achieve the above objectives, performance monitoring ideally should comprise the following functional elements:

- Acquisition and storage of data
- Data analysis and performance identification
- Compensation for degraded performance

11.2.1 Acquisition and Storage of Data

In order to monitor the performance of a system, it is necessary to establish which parameters of the system need to be measured and where and how to obtain these values and how to store them. These parameters may generally be obtained by direct or indirect measurement. A performance monitoring technique may require the measurement of one or more parameters at varying points in time with the data being interpreted via one or more methods of analysis. The parameters to be measured for use within a performance monitoring system may be representative of a variety of system properties. Some commonly used parameters for process plants, for example, are summarized as follows:

- Flow rate
- Pressure
- Temperature
- Concentration
- Liquid level

These measurements are either immediately measurable or dependent on quality analysis such as concentration and should be stored in a structured way in which data can be easily accessed and retrieved. The effectiveness of the storage and retrieval will have a direct influence on the effectiveness of the overall performance monitoring function.

11.2.2 Data Analysis and Performance Identification

Collected data are then used for analysis and diagnosis. The analysis of data in order to draw conclusions about the performance of the system will involve the application of algorithms, formula, or some form of extrapolation "rules". Data analysis may include tha following:

- Data reduction, or the collected data filtering or sampling prior to subsequent processing
- Performance of initial analytic checks on data quality, or the verification and validation of data correctness
- The derivation of indirect performance parameters from collected data

Performance identification may be thought of in the formal sense of diagnosis, where the cause of a known fault with part of plant is determined or in the weaker sense where the overall performance of a system or a subsystem is determined. The activities that comprise diagnosis may be considered as a more detailed investigation of the data to establish the following:

- The underlying cause of poor performance
- The predicted consequences of poor performance
- The recommended maintenance action in response to the poor performance

Results of data analysis and performance identification require that the information presented is in a form that is readily understandable by the maintenance personnel. This output may be fed into the maintenance management system to inform maintenance decision making.

11.2.3 Categories of Performance Monitoring

Monitoring the performance of a system or of its individual components can give a good indication of its condition. This monitoring may involve an ongoing comparison of two or more factors, such as the flow and temperature change across a heat exchanger, or the delivery pressure and running speed of a pump. Performance monitoring can be categorized into steady-state performance monitoring and transient performance monitoring (Yang 2005).

Steady-state performance monitoring is the most widely used form of performance monitoring whereby the system performance parameters are measured during steady operation conditions. Care is necessary to ensure that the system has actually reached a steady condition. In practical applications, the performance parameters of a system are recorded for the healthy condition, normally after commissioning and subsequently, if appropriate, after each significant overhaul has been done. These records form the basis for the evaluation of performance parameters.

There are many cases where the most sensitive monitoring of the steady state of a system can fail to detect an existing problem. In some such cases, it has been found that changes in parameters, which occur during load or speed transients, do provide indicators that allow fault detection and/or fault identification. Transient performance monitoring may be applied to any system, in which parameters can be measured and recorded for comparison. Transient performance monitoring is not only best suited where transient operation regularly occurs, but it also will have complementary value when used with steady-state performance monitoring.

11.3 Performance Monitoring of Control Systems

11.3.1 General Guidelines of Control Performance Monitoring

Control performance usually refers to how well a controller and the process it controls work together. Performance monitoring of control systems should answer the following three questions:

- How healthy is the control system? This is mainly concerned with quantification of control performance. Usually, some measures – whether achievable or ideal and subjective or objective – is used to determine control performance. Some of these measures are the steady-state offset, Integrated Absolute Error (IAE), Integrated Squared Error (ISE), and Mean Squared Error (MSE). A performance measure is used to decide if the performance of a control system is satisfactory or not.
- Why is the control system in a poor health? When the control performance is inadequate, it is important to identify whether a failure in control software or in a control device such as sensors or actuators is the root cause of the poor performance.
- How can control performance be restored to the desired performance level once the degraded control performance occurs?

One principle behind the performance monitoring of control systems is to compare the existing control performance with the best achievable performance from an ideal controller, such as a Minimum Variance Controller (MVC) with a perfectly known plant and a disturbance model or a best tuned PI Controller, to see if the existing performance is close enough to the best performance. If it is not, retuning or redesign of the controller is required. The challenge is to determine incipient problems before the variability in the plant becomes excessive. This predictive fault detection can add real value to industrial operations. Therefore, the real power of the performance monitoring of control systems is to alert operators and management to problems before they are really noticeable.

Two stages are normally taken in the performance monitoring for control systems (Yang et al. 2003b):

- Preliminary stage: Gathering statistical information about the control systems, including (a) percentage of time in service, *i.e.* whether the control system is in service; (b) percentage of correct running period; (c) % in compliance, *i.e.* whether a controlled variable deviates significantly from either a setpoint or Min/Max limits. Identifying the control system, which has some suspect behaviour, in terms of the statistical information gathered.

- Full stage: Applying the existing performance monitoring methods such as Minimum Variances Control to the suspected subsystem identified in the preliminary stage.

Two guidelines are often applied in the implementation of performance monitoring for control systems:

- The cause of poor performance should not be limited to controller design and tuning; other elements in the control systems, such as sensors and actuators, are often responsible for the poor performance. Information from sources other than the controlled variables should be used simultaneously to identify the root cause.
- Cross-plant trend comparison may be crucial to the performance monitoring for control systems. For instance, if a particular control system performs similarly in three out of four plants or devices, something abnormal may be occurring in the fourth control system.

11.3.2 Control Performance Index and General Likelihood Test

11.3.2.1 Control Performance Index

A number of indexes have been used to measure control performance. These include, amongst others, the IAE and the MSE. Here, we introduce a control performance index η, which is defined as the ratio between the actual cost function of the control system and the minimum cost of a desirable benchmark controller.

$$\eta = \frac{J_{\text{benchmark}}}{J_{\text{act}}}, \tag{11.1}$$

where $J_{\text{benchmark}}$ is the cost function of the benchmark controller and J_{act} is the cost function of the actual controller. The cost function can be defined in many ways. A typical cost function previously defined in (9.10) is

$$J(u(k)) = \sum_{k=0}^{\infty} (y^{\text{T}}(k)Qy(k) + u^{\text{T}}(k)Ru(k)),$$

where the weighting matrices Q and R are real symmetric and positive definite, $y(k)$ and $u(k)$ are the system output and controller signal, respectively, and the superscript T denotes the transpose operator.

There are two common used choices for the benchmark controller: Linear Quadratic Gaussian (LQG) controller and Minimum Variance Controller (MVC). The LQG controller is an unconstrained controller and provides the best achievable nominal performance among the class of all stabilizing controllers (Patwardhan et al. 2002). The MVC represents the theoretical lower bound on the achievable

11.3 Performance Monitoring of Control Systems

output variance and requires little process information for the control performance monitoring (Harris et al. 1996). We choose LQG as the benchmark controller to form the control performance index, but MVC can be also used for this purpose. Any cost of the actual controller will be greater than the LQG's cost and the index η takes a value between 0 and 1 ($0 \leq \eta \leq 1$). The index has a value of 1 at the perfect performance and 0 at the worst case. Equation (11.1) becomes (11.2) for the LQG benchmark controller.

$$\eta = \frac{J_{LQG}}{J_{act}} \tag{11.2}$$

11.3.2.2 General Likelihood Test

The LQG controller provides a useful lower bound of the performance achievable by a linear controller when the actual control system is in a healthy state. If both the controllers are applied to the same process, discrepancies between the outputs of the actual controller and the LQG controller will be stable; any fault occurring in the actual controller will make the discrepancies vary significantly (Dai and Yang 2004). A General Likelihood Ratio (GLR) test is employed to monitor the discrepancies in order to detect the faulty controller as early as possible. If we denote the observed actual control signal by $u_{act}(k)$, and the benchmark LQG control signal by $u_{LQG}(k)$, the discrepancies between the two signals $\Delta u(k) = u_{act}(k) - u_{LQG}(k)$ where k is the current time instant, follow a Gaussian distribution with mean μ_0 and variance σ_0 as the number of observations becomes large (Calkins 2005).

The measurement of $\Delta u(k)$ is denoted as $r(k)$ as below:

$$r(k) = \Delta u(k) + r_0(k), \tag{11.3}$$

where r_0 is a white noise following a Gaussian distribution with zero mean.

The GLR test is used to calculate the likelihood ratio s_k of $r(k)$ as (Basseville and Nikiforov 1998)

$$s_k = \frac{\mu_1 - \mu_0}{\sigma_0^2}\left(r(k) - \frac{\mu_0 + \mu_1}{2}\right), \tag{11.4}$$

where μ_1 is the calculated mean of data in a moving observation window, and μ_0 is the previous calculated mean of data in the moving observation window. The mean value μ_i ($i = 0, 1$) is calculated within a moving observation window with a window size K.

$$\mu_1 = \frac{\sum_{k=0}^{K} r(k)}{K}, \tag{11.5}$$

$$\mu_0 \leftarrow \mu_1 \text{ at the next updating instant} \tag{11.6}$$

It might be the case that there is no significant change in the mean value μ, but significant change in the variance σ. Suppose a fault occurs at the instant t_0, before the instant t_0, $r(k)$ has a variance σ_0; after t_0, it has a new variance σ_1. The variance σ_i ($i = 0,1$) is calculated within a moving window with a size N.

$$\sigma_1 = \left(\frac{\sum_{k=1}^{N} (r(k) - \mu)^2}{N} \right)^{1/2} \tag{11.7}$$

$$\sigma_0 \leftarrow \sigma_1 \text{ at the next updating instant} \tag{11.8}$$

The likelihood ratio at the instant k, s_k, is given by (Basseville and Nikiforov 1998)

$$s_k = \ln \frac{\sigma_0}{\sigma_1} \left(\frac{1}{\sigma_0^2} - \frac{1}{\sigma_1^2} \right) \frac{(r(k) - \mu)^2}{2}. \tag{11.9}$$

For both cases, controller faults are identified using a predefined threshold λ as follows:

$s_k > \lambda$: a fault is detected;
$s_k \leqslant \lambda$: no fault is detected and test is continued.

11.3.2.3 Structure for Controller Fault Detection

Structure of controller fault detection based on the above GLR test is shown in Fig. 11.1, which consists of two stages: the residual generation stage and the residual evaluation stage. In the residual generation stage, the benchmark controller, a LQG controller, is designed to run in parallel to the actual controller. The residuals are generated from the difference between the control signals of the actual controller and the LQG controller.

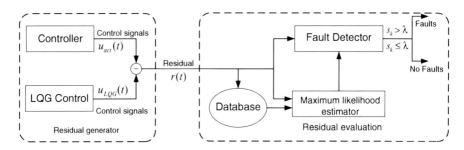

Fig. 11.1 Controller fault detection structure

11.3 Performance Monitoring of Control Systems

The generated residuals are collected and stored in a database for further fault detection. The residual evaluation carries out the GLR test. It consists of three parts: a threshold, a maximum likelihood estimator, and a fault detector. The maximum likelihood estimator computes the mean/variance of the residuals in the moving window. On the basis of the updated mean/variance, the detector tests the observed residual data by comparing whether or not the likelihood ratio is within the predefined threshold range. Once the observed residuals are out of the range, an alarm will show the monitored controller being in a faulty work state. The performance index is another indicator to alert the operators if the performance is degraded.

11.3.3 Performance Compensator Design

In this section, we introduce the design for typical faulty controller performance compensation. A compensator is designed to work together with the faulty controller to stabilize the control process and restore the degraded performance. Consider the feedback control system shown in Fig. 11.2a, in which a model of the controller is $G_c(z)$, and $G_p(z)$ is a plant model. Assume that a fault occurs in the controller and the fault model is described by $G_f(z)$, $m(z)$ is the identified model for the normal open loop, and $m_f(z)$ is the identified model for the faulty open loop.

$$\begin{cases} m(z) \approx G_p(z)G_c(z) \\ m_f(z) \approx G_p(z)G_f(z)G_c(z) \end{cases} \quad (11.10)$$

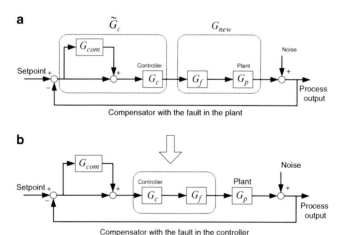

Fig. 11.2 Closed loop with a compensator (Yang et al. 2007)

$G_f(z)$ is calculated as follows:

$$G_f(z) = \frac{m_f(z)}{m(z)}. \tag{11.11}$$

When a controller failure occurs, we can assume that the controller is still in a normal state, the fault part G_f is "shifted" to the plant as shown in Fig. 11.2a, and the equivalent faulty plant model is as follows:

$$G_{new}(z) = G_p(z)G_f(z) \tag{11.12}$$

A virtual controller $\tilde{G}_c(z)$ with a desirable control performance is designed for the equivalent faulty plant $G_{new}(z)$. The compensator $G_{com}(z)$ is designed for the equivalent faulty plant $G_{new}(z)$ to work with the original controller $G_c(z)$ to achieve a reference of control performance set by a virtual controller $\tilde{G}_c(z)$.

$$(1 + G_{com}(z)) \times G_c(z) = \tilde{G}_c(z). \tag{11.13}$$

Therefore, the model of the compensator can be given as follows:

$$G_{com}(z) = \frac{\tilde{G}_c(z) - G_c(z)}{G_c(z)}. \tag{11.14}$$

Here, $\tilde{G}_c(z)$ can be designed in terms of any existing controller design methods or be simply designed as a Proportional-Integral-Derivative (PID) controller. Figure 11.2b shows that the compensator designed for the equivalent faulty plant $G_{new}(z)$ can be directly adopted in the faulty controller to maintain its performance.

11.4 Remote Control Performance Maintenance

11.4.1 Architecture of Remote Maintenance

Control performance can be monitored and maintained in the plant site in most situations. Many SMEs lack resources in technical supports and like to delegate monitoring and maintenance to a service company or software suppliers. In this case, control performance monitoring and maintenance services have to be provided from a remote site. As a result of carrying out remote monitoring and maintenance, the need for any control software supplier's experts to conduct on-site maintenance has been largely eliminated. A typical example is the distributed aircraft maintenance environment (DAME) (DAME 2002), which demonstrated how grid technologies can facilitate the design and development of decision support systems for the

11.4 Remote Control Performance Maintenance

diagnosis and maintenance, in a situation where geographically distributed resources, facilities, and data are combined within a virtual organization.

The remote maintenance system introduced here is different to DAME and is only concerned with control software performance monitoring and maintenance. Also, the communication medium is the public Internet rather than any dedicated network. Collecting data from the real-time process and transferring these data over the Internet make remote monitoring and maintenance much more difficult than traditional local monitoring and maintenance. If not properly dealt with, data transfer over the Internet can cause unacceptable transmission time delay. We adopt the method introduced in Chapter 5 to categorize data into heavy data and light data based on the size of the data. The heavy data are processed in the local site and only the extracted characteristics of the heavy data, such as parameters of the identified models, are transferred to the remote site if necessary. Also, a back-end system at the local site is designed to carry out all the heavy calculations, while a front-end system at the remote site is to allow experts to remotely monitor and maintain the control system.

Figure 11.3 shows the global view of the remote maintenance system. It consists of a local control system, a back-end system, and one or more front-end systems. Process data are collected and processed by the back-end system, in which all the heavy calculations are carried out, such as the control performance index shown in (11.2), the likelihood ratio shown in (11.4) and (11.9), and the compensator model shown in (11.14). The back-end system is also expected to provide communication services to the front-end systems. It uses the full power of object-oriented technology on distributed computing systems to transfer data objects between the remote front-end systems and the back-end system. The front-end systems provide various services over the Internet to the remote users, including a compensator workbench, together with a platform for controller parameter tuning, control performance monitoring, and fault detection.

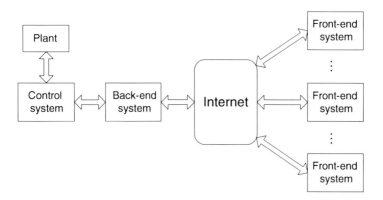

Fig. 11.3 Structure of the remote maintenance system

11.4.2 Implementation of Back-end System

Figure 11.4 shows a typical implementation of a back-end system as shown in Fig. 11.3, which is composed of a data warehouse, process model identification, compensator model identification, benchmark controller, actual and desirable cost functions, performance index, and General Likelihood Ratio test. This implementation is method specific and purely based on the discussion in Sect. 11.3. The data warehouse collects and stores the process variables from the control process. The data warehouse also provides data query and retrieval services for the back-end system components. Each component is driven by the process variables retrieved from the data warehouse. The following steps are involved:

Step 1: The process model is identified based on the process variables stored in the data warehouse. A LQG controller is designed as a benchmark controller. The generated LQG control signal and LQG process output are utilized to compute the LQG cost function.

Step 2: The process variables retrieved from the data warehouse are used to compute the actual cost function. The values of the actual cost function and LQG control function are compared to calculate a performance index value according to (11.1). A poor performance index value will trigger the GLR detector to work.

Step 3: The actual control signals are sent to the GLR detector. The GLR detector traces the discrepancies between the actual control signals retrieved from the data warehouse and the LQG control signals generated by the LQG controller. The likelihood ratio s_k value is computed according to (11.4) or (11.9) and compared with the predefined threshold λ. The result of this comparison determines whether or not any compensation is required. In normal operation, the compensator is inactive. Once s_k becomes greater than the threshold λ, which indicates that the control system is unhealthy, the design of a compensator is activated and sent to the

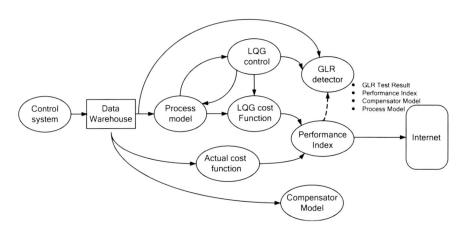

Fig. 11.4 Implementation of a back-end system (Yang et al. 2007)

remote experts for testing and approval. The approved compensator will be returned and implemented in the local control system.

Components in the back-end system are required to cooperate with each other and perform their tasks simultaneously. Multi-thread technologies are employed here to ensure that such cooperation occurs. Each component is implemented in conjunction with a Java thread, which allows the component programs to efficiently perform their tasks independently while simultaneously sharing the latest process variables. The running threads in the components are established with different priorities and managed by a Java thread scheduler. The scheduler can suspend a component by making its thread sleep or kill a component by terminating its thread. It can also monitor all running threads and decide which threads should be running and which should be waiting, according to their priorities and total running time. Other multi-agent technologies are also suitable for the implementation of a back-end system.

11.4.3 Implementation of Front-end System

The front-end system provides a test bed for testing and approval of the compensator and a workbench for the remote tuning of controller parameters, control performance monitoring, and fault detection. The workbench is a Web-based user interface and is illustrated in the case study in next section. This section focuses on the compensator testing and implementation.

11.4.3.1 Compensator Testing

The compensator is tested and tuned in the test bed as shown in Fig. 11.5 by remote experts after it is received by the front-end system. The test bed offline simulates the real implementation of the designed compensator. The compensator and the faulty open-loop model form a virtual control loop. The remote experts monitor the performance of the virtual control loop, tune the compensator parameters, and see how well the compensator works with the faulty plant to follow the setpoint. A small and stable range of deviations between the setpoint and the virtual process

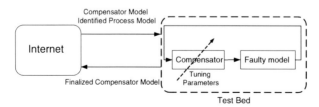

Fig. 11.5 Compensator testing in the remote side

output indicates a healthy performance of the compensator. An unstable and large range of deviations indicate a need for adjusting the compensator parameters. The parameters are manually adjusted until a satisfactory control performance of the virtual control loop is achieved. Only an expert can carry out the compensator testing and parameter tuning properly. For this reason, the test bed for the compensator is placed in the remote front-end system.

11.4.3.2 Compensator Migration

The final version of the compensator is sent back to the back-end system after the testing in the remote side and installed in the local control process over the Internet to restore the degraded control performance. There are two ways to transfer the compensator from the remote side to the local side within the process plant: weak migration and strong migration, as shown in Fig. 11.6. Weak migration transfers only the parameters of the compensator from the remote side to the local side, whereas strong migration transfers both the parameters and the structure of the compensator. Strong migration is employed only if there is no compensator previously installed or the structure of the compensator has been significantly changed since the last implementation. Weak migration is employed to update the parameters of the compensator provided it has the same structure as the compensator tested.

In the implementation described here both migrations are established on the basis of the Java RMI client/server *Remote* interface *java.rmi.Remote*. Weak migration transfers the parameters of the compensator from the remote side to the local side by passing them as the parameters of a remote method invoked. The strong migration, in contrast, transfers the code of the compensator from the remote side to the local side through the Java RMI interface, where the *ClassLoad*( )method is used to search a suitable compensator from the remote front-end system and transfer the code to the local side.

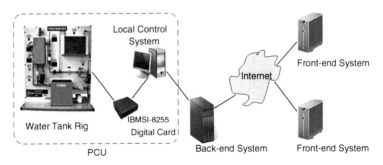

Fig. 11.6 Experimental system layout

11.5 Case Study

11.5.1 System Description

A Process Control Unit (PCU) in the Networks and Control Laboratory at Loughborough University has been chosen for the demonstration and evaluation. The layout of the experimental system is illustrated in Fig. 11.6, which includes the PCU, a back-end system, and a number of identical front-end systems. The PCU has been used in multi-rate controller design in Chap. 8. The local control system and the back-end system are physically located in a same computer and connected with the front-end system via the Internet. Figure 11.7 shows the interface of the remote front-end system. The available performance monitoring list, the available detection list, and the compensation list are displayed in the interface. All the available tools can be added into or removed from the right-hand side windows. The graphs shown in the right-hand side in Fig. 11.7 illustrate the performance index defined in (11.2), the GLR test defined in (11.4), and the test bed window for the compensator defined in (11.14). Other available performance monitoring tools and fault detection tools can be integrated into this front-end system. In Fig. 11.7, two traditional control performance assessment indexes, IAE and MSE, are also implemented in the performance monitoring list. The compensator test bed is illustrated at the

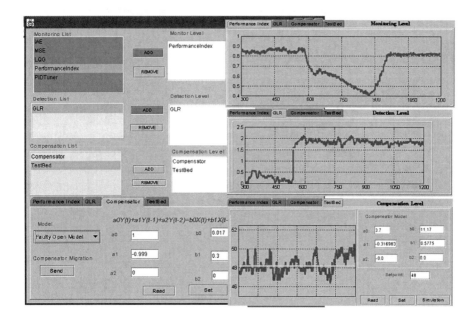

Fig. 11.7 Interface layout of the remote maintenance system (Yang et al. 2007)

bottom part of Fig. 11.7. The remote experts can tune the compensator parameters designed by the back-end system and send the satisfactory set back to the back-end system through the weak or strong migration.

11.5.2 Setting up a Fault

The water tank model is obtained by injecting a step change in the inlet flow rate of the tank and measuring the response in the liquid level of the tank. The sampling interval is chosen as 1 s. The input of the model is the inlet flow rate of the tank, and the output is the liquid level. The identified discrete model of the tank is described as follows:

$$G_p(z) = \frac{0.308z + 0.359}{z^2 - 0.4991z + 0.1}, \quad (11.15)$$

where z denotes the z-transform variable. A LQG benchmark controller is automatically designed in the back-end system for the control performance monitoring and fault detection purposes. The LQG control design is based on the Model Predictive Control (MPC) solution (Patwardhan et al. 2002). Both the prediction and control horizons in the MPC are set as 15 sampling intervals. The cost of the LQG benchmark controller working within the back-end system is compared with the actual control cost made by the actual PID controller; therefore, the performance index is obtained. The differences between the benchmark controller and the actual PID controller are also used in the GLR fault detection.

The original PID controller with the Proportional-Integral-Derivative parameters $K_P = 10, K_I = 0.1, K_D = 0$ is described as follows:

$$G_c(z) = \frac{20.1z - 19.9}{2(z - 1)} \quad (11.16)$$

In order to illustrate how well the remote maintenance system can identify the potential faults occurring in the local control system of the water tank process, a gain, $K = 0.1$, is introduced in the output of the local PID controller at the instant 550 s. As a result, the output of the PID controller is reduced by 0.1 times afterwards. Therefore, the fault model is given as follows:

$$G_f(z) = 0.1 \quad (11.17)$$

Figure 11.8 shows the water level deviating away from the setpoint immediately after the fault is introduced. Figure 11.9 illustrates the performance index significantly dropping from 0.85 at the normal operation to 0.42. The likelihood ratio generated by the GLR test jumps from 0.2 at the normal operation to 1.8 as shown in Fig. 11.10. A faulty state is detected.

11.5 Case Study

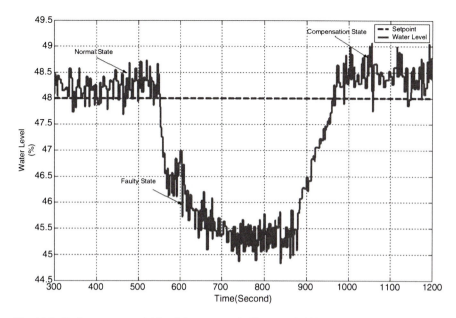

Fig. 11.8 Faulty process variable of the water tank (Yang et al. 2007)

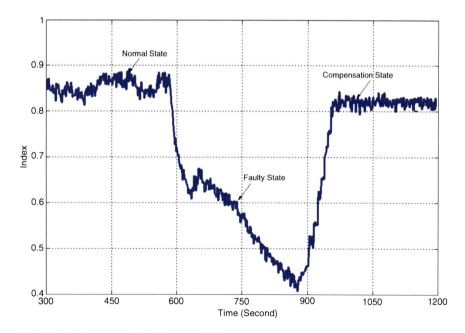

Fig. 11.9 Performance Index (Yang et al. 2007)

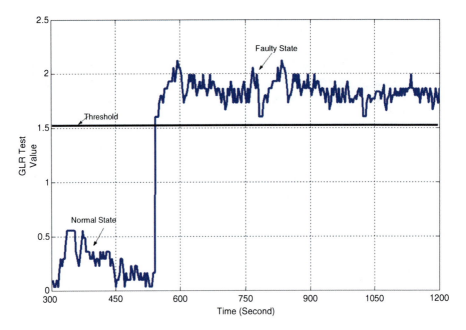

Fig. 11.10 General likelihood ratio (Yang et al. 2007)

11.5.3 Fault Compensation

The desirable virtual controller $\tilde{G}_c(z)$ for the faulty process is set as a PI controller as follows:

$$\tilde{G}_c(z) = \frac{405z - 25.6}{10z - 1} \qquad (11.18)$$

Once the likelihood ratio generated by the GLR test goes over the predefined threshold (1.5 in this study), a compensator is automatically generated in the back-end system for the particular abnormal operation to achieve a reference of control performance set by the virtual controller $\tilde{G}_c(z)$ shown in (11.18). The compensator $G_{\text{com}}(z)$ is obtained by applying (11.18) and (11.16) into (11.14) as follows:

$$G_{\text{com}}(z) = \frac{11.3z - 0.5775}{3.7z - 0.37} \qquad (11.19)$$

The compensator is sent to the front-end system for testing, tuning, and approval in the test bed. The approved compensator is expressed as a linear difference equation as follows:

$$3.7y(k) - 0.37y(k-1) = 11.3x(k) - 0.5775x(k-1), \qquad (11.20)$$

where $y(k)$ and $x(k)$ are the output and input of the compensator at the instant k.

A general compensator is described as the following linear difference equation:

$$a_0 y(k) + a_1 y(k-1) + a_2 y(k-2) + \cdots + a_n y(k-n)$$
$$= b_0 x(k) + b_1 x(k-1) + \cdots + b_n x(k-n), \quad (11.21)$$

where $y(k)$ and $x(k)$ are the output and input of the compensator at the instant k, respectively. The parameters $a_0, a_1, \ldots, a_n$ and $b_0, b_1, \ldots, b_n$ are the compensator coefficients. This predefined general compensator shown in (11.21) has been previously implemented in the local control system with all the parameters being equal to zero. After the compensator for the particular fault is approved in the test bed, the weak migration is employed to remotely install the obtained compensator shown in (11.20) in the real process side. The parameters $\{(3.7, -0.37, 0), (11.3, -0.5775, 0)\}$ are then sent to the predefined general compensator structure. The local operators put the compensator in action if they have been convinced by the simulation results that the compensation is necessary and safe. In this case study, the compensator is put into action at the instant 880 s. Figures 11.8 and 11.9 show that the liquid level of the water tank returns to the desired setpoint and the performance index back to the normal value 0.81 immediately after the compensator being taken place. GLR test indicates that the original controller is still in an unhealthy state as shown in Fig. 11.10, which will remind the local operator to retune the controller during the maintenance stage of the plant.

11.6 Summary

This chapter gives basic concepts and general guidelines of control system performance monitoring. Control performance monitoring method such as control performance index and general likelihood ratio has been introduced as a typical example. A control performance compensator design for a simple scenario is presented for the control system performance maintenance. In order to extend the performance monitoring and maintenance capability into a remote side connected with the Internet, a back-end and front-end system architecture is proposed, which has been demonstrated in a process control unit. In order to avoid the influence of the Internet time delay on the remote operation, all the heavy calculations are carried out in the local back-end system, and only the light data and the parameters of the plant model are transferred to the remote side and dealt in the front-end system.

Another feature of the performance maintenance method introduced in this chapter is to compensate the faulty controller rather than to replace it with a new one. It is obvious that updating a faulty controller might introduce a significant impact on the control system if it is in operation during the updating process. Appending a control signal produced by a compensator into the existing controller will greatly reduce the influence of this maintenance to the normal operation. At the end, the faulty controller can be updated offline once the process plant is shutdown for regular maintenance.

Remote monitoring and maintenance of control system performance might be a good solution for process plant companies with multiple sites in remote locations (for example, power stations and wind electricity generation stations) in order to provide the central support for their geographically dispersed control systems. By using this kind of remote monitoring and maintenance system control software, suppliers can monitor and maintain their control software products remotely over the Internet. The need for any control software supplier's expert to conduct on-site maintenance will be virtually eliminated. Therefore, both time and money can be saved. Furthermore, small and mid-sized companies without internal expertise to maintain their control systems can now rely on the remote monitoring and maintenance system provided by a service company for their system maintenance. This represents a significant cost savings in terms of sharing specialized support staff with other partners. The work in Yang et al. (2003b) gives a case study in an e-support system for Internet-enabled real-time control software, in which control system performance monitoring and maintenance are the main functions. Finally, the generic principle for remote monitoring and maintenance introduced in this chapter can also be applied to fail-safe or backup health monitoring of safety critical systems.

References

Basseville, M., and Nikiforov, I. (1998) *Detection of Abrupt Changes – Theory and Application*, Prentice-Hall, Englewood Cliffs, NJ.

Calkins, K.G., (2005) Introduction to statistics. available at http://www.andrews.edu/~calkins/math/webtexts/stattit.htm

Calvo, I., Marcos, M., Orive, D., and Sarachaga, I., (2006) A methodology based on distributed object-oriented technologies for providing remote access to industrial plants, *Control Engineering Practice*, 14(8), pp. 975–990.

Dai, C., and Yang, S.H., (2004) Maintaining control performance in faulty control systems, *IEEE International Conference on Systems, Man and Cybernetics*, Thissen, W., Wieringa, P., Pantic, M., and Ludema, M. (eds), The Hague, The Netherlands, pp. 5074–5078.

DAME, (2002) DAME Distributed Aircraft Maintenance Environment, available at http://www.cs.york.ac.uk/dame

Harris, T.J., Boudreau, T., and Macgregor, J.F., (1996) Performance assessment of multivariable feedback controllers, *Automatica*, 32, pp. 1503–1517.

HSE, (1995) Safety in the remote diagnosis of manufacturing plant and equipment, *HSE Books*.

Huang, B., and Shah, S.L., (1999) *Performance Assessment of Control Loops: Theory and Applications*, Springer, New York.

Institution of Production Engineers, (1990) *Management Guide to Condition Monitoring in Manufacture*, edited by Davies A.

Jelali, M., (2006) An overview of control performance assessment technology and industrial applications, *Control Engineering Practice*, 14(5), pp. 441–466.

Nougues, A., Vadnais, P., and Snoeren, R., (2002) Performance monitoring for process control and optimization, *ESCAPE–12*, The Hague, Netherlands, pp. 733–738.

References

Patwardhan, R.S., Shah, S.L., and Qi, K.Z., (2002) Assessing the performance of model predictive controllers, The Canadian Journal of Chemical Engineering, 80, October, pp. 954–966.

ScadaOnWeb (2002) http://www.scadaonweb.com.

Sematech, (2002) E-diagnostics guidebook, available at http://www.sematech.org/public/docubase/techrpts.htm.

Thompson, H.A., (2004) Wireless and Internet communications technologies for monitoring and control, *Control Engineering Practice*, 12(6), pp. 781–791.

Yang, S.H., (2005) Remote control and condition monitoring, Chapter 8 in *E-manufacturing: Fundamentals and Applications*, edited by K Cheng, WIT Press, pp. 195–230.

Yang, S.H., Chen, X., and Alty, J.L., (2003) Design issues and implementation of internet based process control, *Control Engineering Practice*, 11(6), pp. 709–720.

Yang, S.H., Chen, X., and Yang, L., (2003) Global Support to Process Plants Over the Internet, *Computer-aided Chemical Engineering*, Chen, B and Westerberg, A.W. (eds), Elsevier, *The 8th International Symposium on Process Systems Engineering*, Kunming, China, 1, pp. 1399–1404.

Yang, S.H., Chen, X, Tan, L., and Yang, L., (2005) Time delay and data loss compensation for Internet-based process control systems, *Transactions of the Institute of Measurement and Control*, 27(2), pp. 103–118.

Yang, S.H., Dai, C., and Knott, R.P., (2007) Remote maintenance of control system performance over the Internet, *Control Engineering Practice*, 15, pp. 533–544.

Chapter 12
Remote Control System Design and Implementation over the Internet

12.1 Introduction

Real-time control systems for industry need regular updating and maintenance. The main reason is that the majority of basic controllers, which have been widely installed, are based on linear system theory and can normally provide a healthy performance only within certain operating ranges even though industrial plants are characterized by a great number of non-linear features. Using the process industry as an example, when raw materials and/or workloads change, the controllers need to be retuned as their parameters are either not sufficiently robust or designed specifically for the original situation. In some cases, it may even be necessary to redesign the controllers, in particular for Advance Process Control (APC) that is more sensitive to changes in circumstance such as changes in raw materials and workloads. Consequently, the related control system software must be modified. Furthermore, the control software is occasionally upgraded to fix software bugs and added-on extra functions. Such actions can only be carried out by the domain experts. However, many small and medium-sized enterprises (SME) do not normally employ such experts because of staffing costs. On the one hand, these SMEs need to ask for support from service companies, and travel by experts and on-site implementation becomes unavoidable. On the other hand, any delay in providing the support may cause significant economic losses.

The Internet and related technologies offer significant new capabilities, which enable experts to remotely monitor the controller performance and the process change, as described in the previous chapter. They also enable remote retuning and redesigning of the control system software. Mathematical models can serve as virtual processes to provide a risk-free alternative to evaluate the new control strategies and/or control parameters. Real-time data collected from the real process can reflect the situation of the current physical process. The benefit of integrating the virtual process and the real process is that the experts working with the virtual process and the operators working with the real process can collaborate at a distance over the Internet and in real time to design, test, and implement the system. For SMEs without any internal experts, the delegated service company and/or control solution supplier can maintain the control system software via the Internet.

It reduces the need for a control software supplier's expert to conduct on-site maintenance (Richardson 2001). The collaborative framework over the Internet has been described as the e-Automation system in Yang and Chen (2002) and the distributed integration framework in Wang et al. (2009).

12.2 Real-time Control System Life Cycle

The life cycle of real-time control systems starts from the system conceptualization, then moves on to design, and then implementation. It moves from the virtual world to the real world. The functions involved include conceptual design, detailed design, verification, implementation, maintenance, and operation. These functions are considered as concurrent functions in order to allow the control system to be updated during on-site operation. Figure 12.1 depicts a generic control system life cycle model presented in Yang et al. (2003a).

The conceptual design is in the conceptualization stage where the boundaries for the considered process are established and the objectives and the specifications are allocated to the control system. The next step in the life cycle is to design a model first for the process and then for the control system. Simulation studies employ computational models, which are implemented by using commercial simulation packages such as MATLAB® or a general programming language such as C, $C++$ or $Java$. The process model obtained should be verified by means of comparing it against the response of the real process, i.e. the real-time data collected from the plant. On the basis of the model obtained, a controller is carefully designed. When the design of the controller is completed, the simulation associated with the process computational model is employed to refine and verify the controller. If the new design of the controller satisfies the specification, it can be smoothly implemented

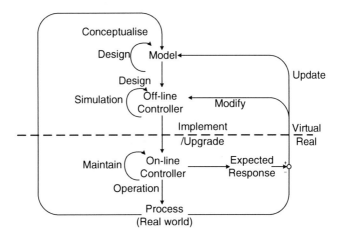

Fig. 12.1 A generic control system life cycle

in the real world. Implementation and/or upgrading cause the real and virtual worlds to interact as shown in Fig. 12.1. The key issue in this step is to minimize the impact on the real process operations of replacing the existing control system with the new design. The online controller needs day-to-day maintenance to operate at the desirable level. Any deviation between the expected and actual responses of the real process will be fed back to the virtual world at the design stage to update the model and modify the controller.

The above life cycle model assumes that control system development is a dynamic and concurrent process and is distinguished by two primary development phases: the virtual world associated with system design and the real world associated with system operation. All the steps in the life cycle are initially carried out in sequence and then are concurrently carried on so that the control system can be upgraded or maintained during the online operation or offline. In practice, the virtual world (the design site) and the real world (the real plant site) are often geographically located in different places, even in different countries. Models and control systems are shipped to the plant site and then installed there with help and advice from the supplier's experts. After the installation and commission, the on-site plant operators will be running the control systems independently and will monitor any deviation between the expected and actual performance of the control systems. These operators still need support from the control system designers once any deviation occurs, which they are unable to deal with.

12.3 Integrated Environments

The Internet and data distributed technologies such as distributed data service (DDS) have made it possible to integrate the virtual and real worlds into a consistent environment without considering their geographically distributed working sites. Therefore, the control system designers from the virtual world can constantly provide support to the on-site operators in the real world. This section introduces the information infrastructure and system design for implementing an integrated environment to enable such interaction between the real world and the virtual world over the Internet.

12.3.1 Interaction Between Real World and Virtual World

The control system life cycle model shown in Fig. 12.1 can be converted into a functional model as shown in Fig. 12.2. Four main components have been identified in the functional model, which are the virtual and real processes and the virtual and real controllers. These four components are linked together either through a local network (Intranet) or the Internet, thus allowing control system designers and

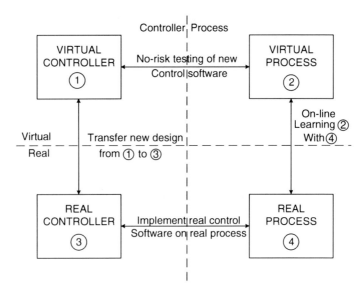

Fig. 12.2 Functional model of the integrated environment

on-site operators to work collaboratively at a distance without considering their geographically distributed working environments. The real controller is located on the plant floor on an on-site computer with the real-time operating system (OS). The real process is a physical plant controlled by the on-site computer. The virtual process is a mathematical model that represents the real-process feature; it can be considered as a mirror of the real process. The virtual and real controllers are considered as the same controller at different stages. While the virtual one is at the design stage, the real one is at the operation stage. There are two boundaries to identify these components; these are virtual vs. real and controller vs. process. From the theoretical point of view, these four components can be located anywhere. However, the real controller and the real process are tightly coupled and need to be located at the same site in order to satisfy the control communication requirement. The virtual controller and the virtual process can be located more or less anywhere over the Internet. A similar model has been used in a virtual control laboratory in Bui et al. (2000), where researchers in the laboratory and operators on the plant floor can work together at a distance and in real time to solve process control problems.

Any two of the four components shown in Fig. 12.2 (from 1 to 4) can be linked together to work for the purpose of designing, testing, and installing a controller in the following four ways:

- A link between the virtual controller and the virtual process (components 1 and 2): Coupling the virtual process with the virtual controller allows designers to perform the testing of new control strategies without any risk to the real plant, to carry out parameter tuning, to select the best dynamic response, and to analyse the sensitivity of the various process variables.

12.3 Integrated Environments

- A link between the real controller and the real process (components 3 and 4): The real controller is implemented in the real process. This link enables the on-site tuning and maintenance.
- A link between the virtual process and the real process (components 2 and 4): This coupling allows designers to identify, update, and improve the virtual process using real-time data collected from the real process. Furthermore, it allows designers to analyse the real process using its virtual counterpart as a tool. This link is particularly meaningful when the real process may change continuously, and such changes must be reflected in its model.
- A link between the virtual controller and the real controller (components 1 and 3): Using this coupling, the new controller can be transferred and/or updated from the virtual side to the real side. In addition, this link can also be used for monitoring the real controller's performance for the purpose of predictive maintenance. It enables the virtual controller to scrutinize the problem and to be developed further.

There are two other possible links: the link between the virtual controller and the real process (components 1 and 4) and the link between the virtual process and the real controller (components 2 and 3). The first one allows designers to test the new design of the controller with the real process. The second one allows the real controller to act on the virtual process. Because the virtual and real components run on different time scales and the network transmission delay is included in the closed loop, it will be difficult and risky to implement them. Thus, safety consideration will prevent these two links in practice. Therefore, we exclude them from the functional model shown in Fig. 12.2.

12.3.2 Available Integrated Frameworks

A software framework provides "the skeleton of an application that can be customized by an application developer" (Johnson 1997). A software framework for remote design, testing, and implementation can be seen as an integrated collaboration environment connected over the Internet with real plants. There are a number of available frameworks aiming at specific application areas. Three of them are summarized here from relevant applications: Service-Oriented Architecture (SOA) based framework, a data-centric framework, and the Java-based Jini architecture.

12.3.2.1 Service-Oriented Architecture

SOA 2008 and Wilkes (2008) present an approach for building distributed systems that deliver application functionality as services to either end-user applications or other services. SOA defines a general application scenario as shown in Fig. 12.3. There are three main protagonists in the architecture, namely the service producer, service consumer, and service registry. The entities participating in a SOA are limited

Fig. 12.3 Service-oriented architecture

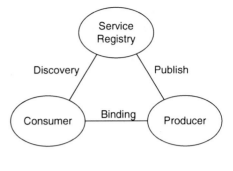

Fig. 12.4 Elements of service-oriented architecture

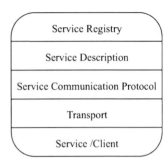

to a single role only for the time of a single interaction. In the scenario, service producers publish their services on one or more service registries. When publishing services, a description is supplied, which is in turn used by the registry to categorize the service. Service consumers use a search facility on the registry to find a desired service. The search can be performed based on different criteria, e.g. information category or functionality. If it has been successful, a set of service descriptions is delivered to the consumer, who then chooses the service most suitable for their intended purpose. With the information contained in the description, a connection can be established to the service endpoint of the producer, i.e. the consumer's client application binds itself to the service endpoint for invocation.

The elements of the functional structure in SOA are illustrated in Fig. 12.4 and described as follows.

Service and client: service describes an actual service that is made available for use. If the service uses other service it becomes a client. Software applications or functional entities are included in a service layer.

Transport: transport is the mechanism used to move service requests from the service consumer to the service provider and to move service responses from the service provider to the service consumer.

Service communication protocol: service communication protocol is an agreed mechanism, which the service provider and the service consumer use to communicate what is being requested and what is being returned.

12.3 Integrated Environments

Service description: service description is an agreed schema for describing what the service is, how it should be invoked, and what data are required to invoke the service successfully.

Service registry: service registry is a repository of service and data descriptions, which may be used by service producers to publish their services and used by service consumers to discover or find available services. The service registry may provide other functions to services that require a centralized repository.

The most important attributes of SOA are high levels of agility and flexibility. By combining a number of small services and focusing on a distributed architecture, development processes are supposed to be more efficient and cost-effective. The evolutionary approach of building SOA results in federated distributed infrastructures suitable for and easily adaptable to remote design, testing, and implementation for real-time control systems. SOA reduces the risk of erroneous trends, software migration, complex component integration, and can also reduce development times. SOA enables a better communication between the different teams in a developed project/system (Ermolayev et al. 2004).

12.3.2.2 Data-Centric Framework

Data-centric frameworks (Chand & Felber 2004) are being developed for distributed database applications as a result of the recent availability of high performance messaging and database technologies and, to a certain extent, as a result of the increasing adoption of SOA and Web Services in software applications. The data-centric model is characterized by distributed participants together with data-centric interactions between these participants and the use of the publish/subscribe model for large volumes of data.

The data publish/subscribe model is shown in Fig. 12.5. Publish–subscribe middleware is the key enabling technology for data-centric design. In contrast to the "central server with the many clients" model of enterprise middleware, publish–subscribe nodes simply subscribe to data they need and publish the information they produce. Thus, the data dictionary map is directly translatable to publishers and subscribers. The middleware passes messages between the communicating nodes.

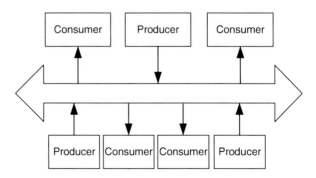

Fig. 12.5 Data publish/subscribe model

The fundamental communications model implies both discovery – what data should be sent where, and delivery – when and how to send it. Several middleware technologies and standards have been applied to the construction of distributed systems including the Java Message Service (JMS) and the Data Distribution Service (DDS). Comparing with JMS, DDS can explicitly control the latency and efficient use of network resources, which is a critical issue in real-time applications such as remote testing and implementation over the Internet (http://www.omg.org/technology/documents/formal/data_distribution.htm; http://www.rti.com/products/data_distribution/index.html).

Communications and data management for data-centric framework are implemented by middleware with no intervention from the users of the framework. A typical commercial middleware for this purpose is Data Distribution Service from Real-Time Innovations (RTI) (http://www.rti.com/products/data_distribution/index.html).

12.3.2.3 Java-based Jini Architecture

The Java-based Jini technology (Ghiassi 1999) has been introduced to enhance interoperability of computer systems and to establish a dynamic communication channel between distributed software components. Jini's objective is to create a network-centric computing environment. In this environment, Java-based applications serve as the portability glue and interoperability environment, and the Jini technology replaces the central distributed system with a federated distributed system. Within a Jini system, users are able to share services and resources over a network.

A Jini system is a federation of many components and consists of the following parts:

- A set of components that provides an infrastructure for federating services in a distributed system
- A programming model that supports and encourages the production of reliable distributed services
- Services that offer functionality to any other member of the federation

The Jini infrastructure, as shown in Fig. 12.6, works using lookup and registry services, with lookup being its main function. This service allows devices, including hardware devices and software programs, to advertise their services and to communicate with one another. Each device on this network is considered as an object, which can join a lookup service using a pair of protocols called discovery and join. An object first uses the discovery service to locate a lookup service and then uses the join protocol to join the service.

Once a device is plugged into a network, it first tries to locate a lookup service by multicasting a request on the local network for any lookup service to identify itself. Once a lookup service is located, the object (device) can register itself with the lookup service by providing it with a set of attributes. These attributes represent

12.3 Integrated Environments

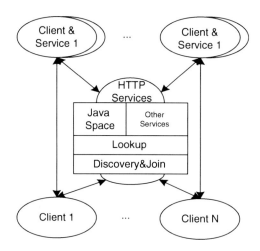

Fig. 12.6 Jini infrastructure

a set of parameters describing the service that the object can offer. Finally, the lookup service serves as the repository for all the services.

The architecture of the Jini system is designed to support the dynamic configuration of network objects. Those objects registered with the network can be moved from one location to another; they can also be detached from the network easily without impacting the operations of the network itself.

12.3.3 Architecture of a General Integrated Environment

A general integrated environment for remote design, testing, and implementation of real-time control systems should have an appropriate federated information infrastructure. The following features should be met by such a general integrated environment:

- The environment must be implemented using an open architecture based on mainstream computer technologies and non-proprietary standards
- The environment must provide multiple aspect services such as remote design, testing, and implementation
- The environment must provide an infrastructure for federating participants, in which each participant can join and detach from the infrastructure without impacting upon the operation of the environment itself
- The environment must support the collaboration among the participants and provide the right amount of information for it

Figure 12.7 illustrates a general architecture for the required integrated environment, in which all the functions are implemented as various services and communications between services are illustrated via service bus. The service bus might

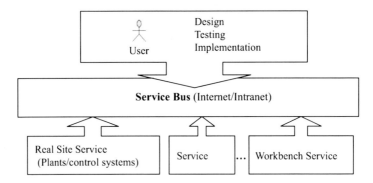

Fig. 12.7 General integrated environment architecture

be the Internet or intranet. The real world shown in Fig. 12.2 is represented as the real-site service in Fig. 12.7. The workbench service provides a place for assembling, testing, and implementing the available software components designed in geographically dispersed locations. Collaborations between different services also take place in the workbench (Yang et al. 2003a). One or more users of the integrated environment such as control system designers and on-site operators invoke the services provided by the integrated environment.

Concerning real-time data transfer, all data are wrapped in XML or HDF format (see Chap. 5 for details) and then published/subscribed through DDS or JMS middleware.

To deal with collaboration issues and possible conflicts between the various design teams or various services, an efficient notification mechanism is required, which can be achieved, for example, by publishing/subscribing messages over the Internet if the architecture follows the data publish/subscribe model shown in Fig. 12.5. Services can publish collaboration or conflict messages and other services can be informed by subscribing interested messages. Therefore, different services can work together in real time.

The above general architecture can be implemented through the SOA combined with a data-centric model, i.e. data publish/subscribe model. It can also be implemented in other ways. In the next section, we discuss a typical implementation of the general integrated environment using the Jini technology.

12.4 A Typical Implementation of the General Integrated Environment

An implementation of a Jini-based integrated environment is illustrated in Fig. 12.8, which is a flat structure. The structure is composed of three basic elements wired by the service bus: a Jini server and a web server, a real site (the real world), and a design workbench (the virtual world). The web service provides a Graphic User

12.4 A Typical Implementation of the General Integrated Environment

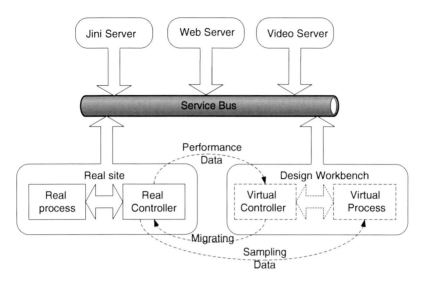

Fig. 12.8 Jini-based integrated environment

Interface (GUI)-based service and is used to retrieve the information embedded in the Jini lookup service. The Jini service provides back-end management functions. Since the Jini service and the web server are tightly coupled, it is best if that are co-located.

The function of a Jini server is to establish and manage the communication channel rather than pass the information between the service providers and the service consumers. Once the connection is established, the assigned developing actions can be carried out by individual elements. The individual elements are called services here. At the real site, the location of the real controller and the connection between the real controllers and the real process are prior-specified and static. In contrast, at the virtual site, the instances of the workbench are dynamically created and there could be more than one instance from the same workbench, which depends on the design tasks assigned. The integrated environment logically works through several service pairs between the real site and the virtual site, including migrating the designed controller from the virtual site to the real site, transferring performance and sample data between them, and establishing the connection between the components. These service pairs are established through the links via the service bus, which includes not only the Jini protocol, but also other protocols such as HTTP.

There are a number of advantages in using the Jini technology to implement the integrated environment. One of them is that the Jini infrastructure provides a flexible solution for distributed communication and collaboration. In the functional model shown in Fig. 12.2, a large number of small services such as controller design and model updating are independently provided from geographically dispersed locations. They need at times to communicate with each other for a short or long period. The Jini infrastructure allows the services to join or leave the network without the need to reconfigure the network. In addition, the Jini lookup service acts as the system service repository and provides the system with the service management. For example, remote designers should redesign the controller and/or tune the controller parameters once the performance of the controller fails, for some reasons, to meet the design specification. When a new controller or new parameters are available, the remote designers register them in the Jini lookup server and inform the on-site operators to download them from the lookup server at the proper time rather than directly download them from the designer's server into the on-site controller. The latter is not allowed for security reasons.

12.4.1 Design Workbench

Suppose that the virtual process (the model) and the virtual controller have been designed in geographically dispersed locations. The workbench is the virtual place in the network for loading the designed model and controller from their designer's sites, synchronizing the simulation of the model and controller and registering them in the Jini Server after successful testing. A conceptual structure of the workbench is illustrated in Fig. 12.9. For the loading operation, the workbench extends the ClassLoader class, a default Java class, to facilitate the dynamic loading of the model and controller from their designer's sites. The synchronization of the execution of the designed model and controller can be easily implemented by taking advantage of Java's multiple thread capabilities. Standard Jini packages are used for service registration and subscription (Ghiassi 1999).

12.4.1.1 Running the Workbench in "Fat" and "Thin" Styles

Testing a controller and a model in the workbench over the Internet can take place in a "fat" client or a "thin" client style. Here, the client means the designer site or the workbench, the server means the Jini and Web server. The "fat" client indicates

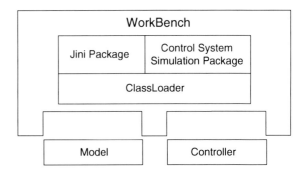

Fig. 12.9 Workbench conceptual structure

12.4 A Typical Implementation of the General Integrated Environment

that there is nothing left to do on the server side; the service runs entirely within the client side. The role of the Jini server is to register the service with the service locator, a standard Jini function, and stay alive so that it can send and receive the message to and from the client to carry out collaboration among multi-clients. A typical class diagram for the Virtual Machine (VM) in the "fat" client style is shown in Fig. 12.10, in which a client asks for an IModel interface implementation, a mathematical model for a controller to be evaluated, and receives an ImodelImpl, the full implementation of the IModel interface. The IModelImpl runs entirely within the client and does not need to communicate with the server at all.

In contrast, the "thin" client leaves some processing tasks on the server side. A proxy exists on the client side to take calls from the client, invoke the methods on the server side, and return the results to the client. A general class diagram of the VM in the "thin" client style is illustrated in Fig. 12.11 with the same functionality as that shown in Fig. 12.10. A client asks for an IModel interface implementation and then obtains an IModel implementation stub (IModelImpl_Stub), a proxy of the implementation of the IModel, rather than the full implementation of the IModel interface. The IModelImpl_Stub receives the request from the client and forwards it to the server and then carries out the main process on the server side. The results will be sent back to the IModelImpl_Stub and then to the client.

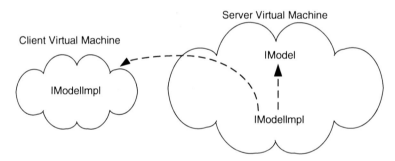

Fig. 12.10 "Fat" client style

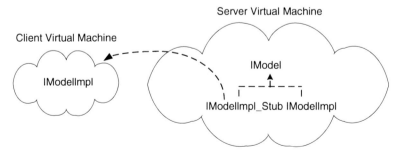

Fig. 12.11 "Thin" client style

Considering that a controller must meet the real-time requirements and there is a potential large time delay on the Internet, it is improper to locate the controller away from the process model over the Internet. Therefore, the "fat" client style would be a wise choice for the controller as it and the model are physically located in a same place. For the model design, the choice will depend on the model size and the computing resource required. For example, a complicated model may require a large amount of computing resource, which is difficult to provide on the client side and easy on the server side. Therefore, the "thin" client style should be selected for a more complicated model. In comparisons, the "fat" client is easier to implement and manage in multiple user environments. The downloading time in the "fat" client style is longer than in the "thin" client style. In this chapter, both the controller and the model are in the "fat" client style because the model size is small and the required computing resources are easily met.

12.4.1.2 Testing a Model and a Controller at the Workbench

The sequence of testing a model and a controller at the workbench and then registering them in the Jini server is illustrated in Fig. 12.12 in the Unified Modeling Language (UML) format. The whole sequence consists of six stages as follows.

Stage 1. A workbench web page is requested by a designer and retrieved from the web server.
Stage 2. The workbench loads the designed controller from the local file system.
Stage 3. A testing model (a virtual process) is selected from the Jini server. Since there may be several available models, the user needs to select the correct one.
Stage 4. The designed controller and the testing model are bound together to build a closed-loop control system within the workbench.
Stage 5. The controller is assessed by running the simulation of the closed-loop control system at the workbench. If the designer is satisfied with the performance of the closed-loop system, then proceed to Stage 6; otherwise, the controller should be modified or redesigned and tested again from Stage 1.
Stage 6. The new controller is registered to the Jini server and its readiness for use at the real site is announced.

12.4.2 Implementing a New Design of a Controller

Any on-site control system will have certain safety-critical requirements. It is unlikely to allow the updating of the existing controller with the new controller by the controller designer directly from a remote location under any circumstances. The practical solution is to let the on-site operators replace the existing controller with the new design under the one-line guidance of the controller designer.

12.4 A Typical Implementation of the General Integrated Environment

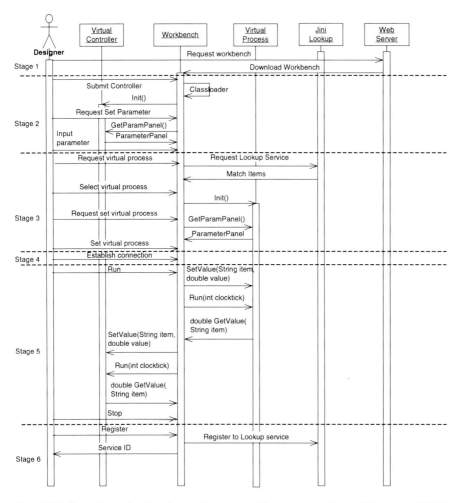

Fig. 12.12 Procedures of testing the model and controller at the workbench (Yang et al. 2003a)

The flexible communication functions provided by the Jini technologies facilitate such a controller updating operation.

Figure 12.13 shows a procedure for updating a controller at the real site, which is described as the following four stages:

- Loading stage. In the loading stage, the main action is to prepare a new controller, which includes selecting a controller from an available list, loading the controller, and establishing the connection between the controller and the process as illustrated in Fig. 12.13a
- Activating stage. When the new controller is ready, the activating action enables the new controller to participate in the existing control system, but the

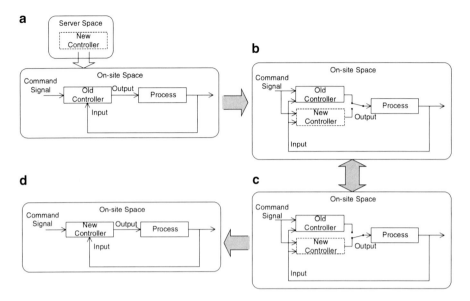

Fig. 12.13 Controller updating; (**a**) loading; (**b**) activating; (**c**) switching; (**d**) updating (Yang et al. 2003a)

output of the new controller does not affect the real process; the system is transferred from the one shown in Fig. 12.13a to the one in Fig. 12.13b
- Switching stage. The performance of the new controller is evaluated by comparing the outputs from the two parallel controllers, i.e. the new and existing ones. After this test, the existing on-site controller and the new controller can be swapped, and the system is then transferred from Figs. 12.13b, c. Further evaluation is carried out there
- Updating stage. Once the on-site operators are satisfied with the new controller, the updating action will complete and terminate the updating operation, which is shown as Fig. 12.13d. The effect on the controlled process must be minimized as much as possible during the updating; otherwise, a disturbance will be introduced into the operation and the production will be affected

The interaction among the on-site operators, the real control system, the Jini server, and the new controller during the updating is shown in Fig. 12.14, which demonstrates the typical event-notifying mechanism. In order to respond to the information (called the event here) that the new controller is available, the on-site control system needs to be first registered in the Jini server with the event listener. Once the new controller is registered, the Jini server will inform the registered on-site control system that the new controller is available. On the basis of the information provided, the on-site operators will decide whether to install the new controller or not. Should they decide to do so, the above four stages updating operation will be carried out.

12.4 A Typical Implementation of the General Integrated Environment

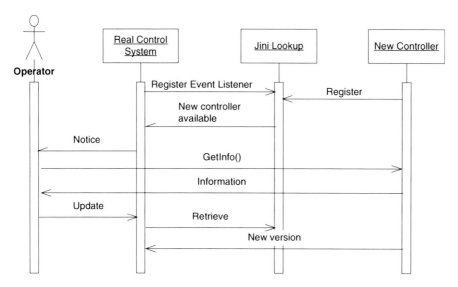

Fig. 12.14 Interaction in the controller updating process (Yang et al. 2003a)

12.4.3 Collaboration in the Integrated Environment

The collaboration requirements in the integrated environment across the virtual and real worlds are as follows:

- The real-site operators need to know whether the newly designed controllers are available or not
- The controller designers need to know whether the on-site controller has a satisfactory performance or not, and need to be informed once the model is updated
- The model designers should prepare to update the model once a new set of real-time data have been collected by the on-site operators

Collaboration in the integrated environment among the design workbenches, the real site, and the Jini server is partly implemented through the service pairs that provide the services at the one end and consume the services at the other end. A lightweight asynchronous mechanism is necessary to provide an efficient way for the controller and model designers and the real-site operators to work as a team over the Internet. The distributed message event mechanism supported by the Jini infrastructure is used here for building this collaboration. The Jini infrastructure provides the basic classes to facilitate the distributed event mechanism, which consists of a pair of event classes: *EventRegistration* and *RemoteEventListener*. This mechanism is fully aware of the network status such as network failure and wrong order of the event delivery. Similar to the event-handling system in the Java Swing package (Walrath and Campione 2000), several pairs of event classes are designed based on the Jini distributed event mechanism.

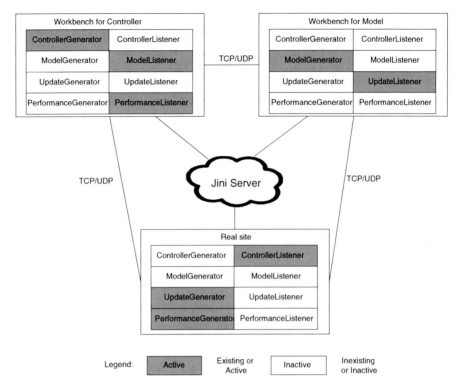

Fig. 12.15 Service producers and consumers in the integrated environment

As shown in Fig. 12.15, any pair of event classes generally includes two parts: a generator for producing an event and a listener for dealing with the event. The highlighted classes in Fig. 12.15 illustrate their locations in the integrated environment. For example, the classes *ControllerGenerator*, *ModelListener*, and *PerformanceListener* are located at the workbench for the controller; the classes *ModelGenerator* and *UpdateListener* are at the workbench for the model; the classes *ControllerListener*, *UpdateGenerator*, and *PerformanceGenerator* are at the real site. The *ControllerGenerator* generates an event to announce that a new designed controller is available; the ControllerListner at the real site will deal with this event. The *ModelGenerator* produces an event to announce that a new version model is available; the *ModelListener* will receive the message and carry out the relative actions. The model designers at the workbench for the model may notice that a new set of real-time data is available through the *UpdateListener* and *UpdateGenerator* class pair. The PerformanceListener at the workbench for controller deals with the event generated by the PerformanceGenerator. This event is used to inform the controller designers whether the performance of the on-site control system is satisfactory or not.

All communication goes through the Jini server. The HTTP protocol and TCP Java sockets have been used to build the communication channels. For example, the

TCP and UDP sockets are used for transferring data from the real site to the design workbenches, and the HTTP protocol is used for downloading the design workbench from the web server.

12.5 Case Study

In order to demonstrate and evaluate the integrated environment described above, the Internet-based control system for the water tank in our process control laboratory (Yang et al. 2003a, b) has been chosen as a test bed. The layout of the test bed is shown in Fig. 12.16. Comparing with the test bed used in the Virtual Supervision Parameter Control (VSPC) shown in Fig. 7.21, there are three differences: (a) a local control system is implemented with the water tank; (b) a Jini server is added with the Web server; and (c) remote operators (client1 to clientM) for VSPC have been replaced by remote designers. Otherwise Fig. 7.21 could have an identical hardware structure to that of Fig. 12.16 here.

The local control system and the DAQ instrument are connected through a RS-232c serial cable and exchange real-time data between them. The local control system is connected to the Web & Jini server. The Web & Jini server provides the Jini services as well as the Internet services and also establishes the connections between the clients and the local control system. Several remotely located designers (users) are allowed to simultaneously monitor the water tank using this Internet-based control system over the Internet.

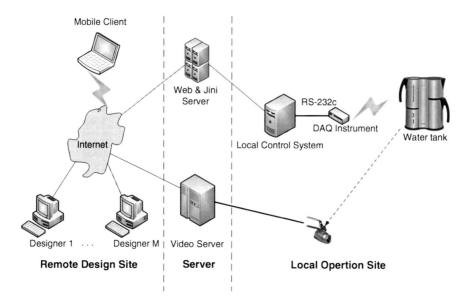

Fig. 12.16 Hardware structure of the test bed

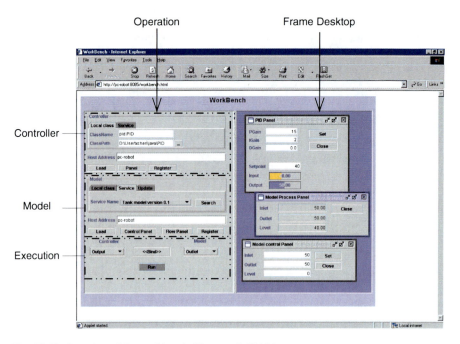

Fig. 12.17 Snapshot of the workbench (Yang et al. 2003a)

12.5.1 Workbench for Testing

As shown in Fig. 12.17, the workbench is divided into two parts. The left-hand side is the operation panel for carrying out the services; the right-hand side is the frame desktop for organizing the frames generated by loading a model and a controller and binding them for testing. Figure 12.17 illustrates loading a controller from the local machine, retrieving a model stored on the Internet, binding them together, and carrying out the simulation of the closed-loop control system at the workbench for the model and controller testing.

12.5.2 Testing the Model and the Controller of the Water Tank at the Workbench

The default on-site controller on the plant floor is a PID controller without the integral action limit; the new design of the controller in the virtual world is a PID controller with the integral action limit (Marlin 2000). The control formula is omitted here for the sake of simplicity. The water tank shown in Fig. 12.16 is a pure capacity process and can be described as follows:

$$y(s) = \frac{K}{s} x(s) \qquad (12.1)$$

where $y(s)$ is the liquid level in the water tank, $x(s)$ is the open percentage of the outlet valve, K is the gain of the capacity process, and s is the Laplace transfer symbol. The initial value of K is calculated based on the geometric theoretical value of the tank and is updated according to the real-time data from the plant as follows: When a new set of real-time data are collected by the on-site operators and are ready for use, the class *UpdateGenerator* will generate an event to announce the availability of this set of data in the Jini server, and the class *ModelGenerator* at the workbench for model receives the announcement and automatically instructs the model designers to update the existing model. This collaboration process follows the sequences described in Sect. 12.4.3.

Figure 12.18 shows a comparison between the responses of the original and modified models and the actual output of the process. It is generated by a sudden increase in the opening of the inlet hand valve at the time 35 s and changing back to the original position after about 30 s at the time 65 s (see the period between 35 and 65 s in Fig. 12.18). The result shows that the modified model fits the real process better and is a great improvement over the original model. Since the model is quite simple, the model adopts the "fat" client structure shown in Fig. 12.10.

Ideally, the real-time data should be provided on a real-time basis, so that the model can always keep track of the real process. However, unexpected noise can mislead the model; the continuously updating model would cost a lot of network resources and in any case is unnecessary. A practical way adopted here is to provide a batch set of real-time data. If the operators observe any change in the process, they can record the change in a data file. When the data file is ready, the model designers will be instructed automatically by the collaboration mechanism in the integrated

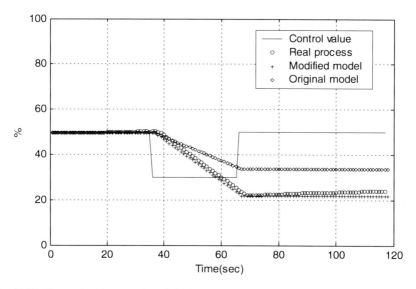

Fig. 12.18 Comparison between the original and updated models and the actual process behavior (Yang et al. 2003a)

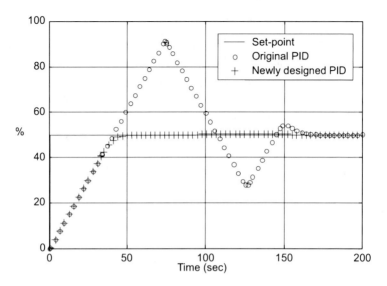

Fig. 12.19 Comparison of the newly designed and the original PID controllers (Yang et al. 2003a)

environment to update the model. It will reduce the network traffic load and will more easily coordinate the model designer's work.

The new design of the controller is compared with the original one at the workbench for testing. The updating model of the water tank is employed for both of the controllers. The simulation result is shown in Fig. 12.19. Obviously, under the same setpoint change, the new design has less overshoot and better performance compared with the original design. This is because an integral action has been added in the design of the updated controller.

12.5.3 Installation of the New Design of Real Controllers

The evolution of updating an existing controller at the real site as shown in Fig. 12.13 has been implemented as shown in Fig. 12.20, where the four snapshots correspond to the four steps in the control system updating. These four steps are loading, activating, switching, and updating. After these four steps, the final system is expected to have the same structure as the original one, but with the new design for the controller replacing the original one. This implementation can and should be repeatedly carried out during the control system life cycle.

Figure 12.21 illustrates the dynamic trends of a set of control system variables during the four steps of the new controller installation. The whole installation is smoothly carried out and there is hardly any impact on the ordinary process operation. As mentioned above, the integral action limit is added to the new controller. When the setpoint remains unchanged, the result shows that the two

12.5 Case Study

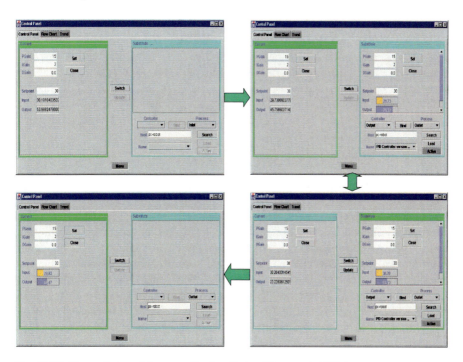

Fig. 12.20 The evolution of the controller installation (Yang et al. 2003a)

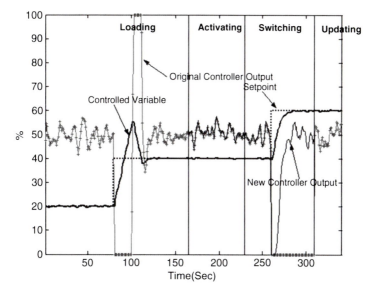

Fig. 12.21 Control system variables during on-site implementation (Yang et al. 2003a)

controllers have the similar output values. Once the step change is introduced on the setpoint, the new controller demonstrates less overshoot in the controlled variable as shown in the period between 250 and 275 s compared with the variable controlled by the original controller in the period between 75 and 125 s in Fig. 12.21. This is because the new design of the controller eliminates the integral action when its output hits the limit. In the updating stage, the original controller is completely replaced by the new one. The substituting point is clearly shown in Fig. 12.21 at around 315 s.

12.6 Summary

This chapter illustrates a way of remotely designing, testing, and updating real-time control software through the Internet in an integrated environment. This integrated environment implements the functionality of control system design, testing, and installation over the Internet by using the Jini, SOA, and DDS technologies even though only the Jini-based design and implementation has been given in the chapter. It covers all the stages in the generic control system's life cycle, from the virtual world to the real world, and from design to implementation. As a consequence, the model and controller designers and the on-site operators can work together through the integrated environment, as a team from geographically dispersed locations, to provide maintenance support to all the authorized industrial processes. It virtually eliminates the need for any control software supplier's expert to conduct on-site maintenance. Another benefit is that small and mid-sized companies without the internal expertise can buy specialized support staff from a service company for their system maintenance and therefore reduce costs.

There is still work to be done in putting the integrated environment into practice. At the moment, the controller and the process model are designed outside the integrated environment and loaded into the workbench in the virtual world for testing purposes. It is helpful to include the initial design of both the controller and the model in the integrated environment. A practical approach is to integrate some commercial design packages such as MATLAB® or SIMULINK® into the environment. The challenge is that the commercial design packages must be open to users; hence, the computer language used in the packages must be compatible with Internet-enabled language such as Java.

The integrated environment developed here can be further extended to wider applications such as e-Diagnostics and e-Support for Internet-enabled systems. E-Diagnostics (2002) enables an authorised equipment and software supplier's field service person to access facilities such as equipment and software from outside the enterprises via the Internet or Intranet. Access includes the ability to remotely monitor, diagnose problems or faults, and configure and control the equipment and software in order to bring it rapidly them into a full productive state. The above e-Diagnostic function can be incorporated into the integrated environment if real-time data that represent the state of the on-site equipment and/or software are regularly collected at the real site and transferred to the e-Diagnostic function when required. The challenges are how to identify control system software, which needs to be updated, how to maintain the control system software with a minimum disturbance to the controlled processes, and furthermore how to ensure the safety and security of the Internet-based operations. Nevertheless, this chapter presents a new area for control engineers and software engineers in real-time software application.

References

Bui, R.T., Perron, J., and Fillion, C., (2000) Model-based control for industrial processes using a virtual laboratory, *Lecture Notes in Computer Science*, 1821, pp. 671–682.

Chand, R. and Felber, P., (2004) A reliable content-based publish/subscribe system, *In the 23rd International Symposium on Reliable Distributed Systems*, pp. 264–273.

Data Distribution Service for Real-time Systems, available at http://www.omg.org/technology/documents/formal/data_distribution.htm

Ermolayev, V., Keberle, N., Plakin, S., Kononenko, O., and Terziyan, V., (2004) Towards a framework for agent-enabled semantic web services composition, *International Journal of Web Services Research*, 1(3), pp. 63–87.

IBM SOA Foundation, (2008) Aerospace and defense, available at http://www-306.ibm.com/software/info/developer/solutions/asad/index.jsp.

Ghiassi, M., (1999) Applying Java technology to manufacturing integration and automation, *International Journal of Flexible Automation and Integrated Manufacturing*, 7, pp. 355–377.

Johnson, R.E. (1997), Frameworks = Components + Patterns, *Communications of the ACM*, 40 (10), pp. 39–42.

Marlin, T.E., (2000) Process control: designing processes and control systems for dynamic performance, McGraw-Hill, New York.

Richardson, T., (2001) When maintenance meets the Internet, *A-B Journal*, September, pp. 25–27.

RTI Data Distribution Service, available at http://www.rti.com/products/data_distribution/index.html

Sematech, (2002) e-Diagnostics guidebook, available at http://www.sematech.org/public/documbase/techrpts.htm

Walrath, K. and Campione, M., (2000) The JFC swing tutorial: a guide to constructing GUIs, Addison-Wesley, Boston MA.

Wang, Y.H., Yang, S.H., Grigg, A., and Johnson, J., (2009) A DDS based framework for remote integration over the Internet, *Proceedings of the 7th Annual Conference on Systems Engineering Research*, Loughborough.

Wilkes, S., (2008) SOA-Much More Than Web Services, available at http://dev2dev.bea.com/pub/a/2004/05/soa_wilkes.html

Yang, S.H. and Chen, X., (2002) Web-based solution for control system maintenance, *Proceedings of The 5th Asia-Pacific Conference on Control & Measurement*, China Aviation Industry Press, Dali, China, pp. 250–255.

Yang, S.H., Chen, X., and Yang, L., (2003) Integration of control system design and implementation over the Internet using the Jini technology, *Software Practice & Experience*, 33, pp. 1151–1175.

Yang, L., Chen, X., and Alty, J.L., (2003) Design issues and implementation of Internet-based process control systems', *Control Engineering Practice*, 11(6), pp. 709–720.

Chapter 13
Conclusion

13.1 Summary

Motivated by the rapid development and adoption of the Internet, we consider the design and applications of Internet-based control systems for industry. Our focus is not only on control engineering aspect, but also on computer science while dealing with the various issues relating to the contribution of the Internet to remote closed-loop control systems. The fact is that neither of these two disciplines can solely offer a comprehensive solution to the design issues of Internet-based control systems. Multidisciplinary solutions drawing from both domains appear to offer the most promising path forward.

From this standpoint, we began the book by considering the requirement specification for Internet-based control systems. In Chap. 2, we introduced an ideal integrated distributed architecture, in which the control system is linked with the Internet at all levels in the control system hierarchy. Based on a general structure of networked control systems, three canonical Internet-based control system architectures were proposed in Chap. 3. Chapter 4 answered the question on how information should be displayed to the remote operators as they are some distance from the main control room and only have only very limited media available to them. Transferring a large amount of data over the Internet will never be straightforward and will be particularly difficult if real-time constraints are applied. Chapter 5 provided a solution for real-time data transfer over the Internet. Data have been classified into light and heavy data. XML and HDF have been used to wrap the various data and transfer it as a data object using the Java Document Object Model. Chapters 6 and 7 dealt with Internet-transmission delay and data loss from the computer network perspective and the control theory perspective, respectively. A combined network infrastructure was investigated in Chap. 6, which integrated TCP, UDP, Real-time Control Protocol, and the Network Time Protocol. The advantages of each transmission protocol have been used in the integrated infrastructure with the elimination of their disadvantages. Chapter 7 gave a number of options for overcoming the Internet-transmission delay and data loss from the control theory perspective, including virtual supervision parameter control, model-based predictive display tele-operation, intelligent autonomy control, and control systems with a variable sampling time. The emphasis has been placed on multi-rate control, which deploys a two-level control architecture, the lower level of which is running at a higher frequency to stabilize the plant and guarantee the plant being under control in case a large transmission delay occurs. The higher level of the control architecture implements the global control function and is running at a lower frequency to reduce the communication load and increase the possibility of receiving data on time. Chapters 8 and 9 gave the details of designing SISO and MIMO multi-rate Internet-based control systems, respectively. The key issue is to stabilize the control system with properly selected parameters. A necessary and sufficient criterion for asymptotic stability was given in Chap. 9, but is limited to linear systems. Safety and security checking share more similarities than there are differences between them. Chapter 10 presented a framework for security checking, which is implemented by using the 'What-If' approach, which was originally designed for safety checking. An extended process control event diagram (PCED) was used as a starting point and a set of new attributes and "guidewords" were applied in the safety checking. Chapters 11 and 12 applied the Internet-based control systems in remote control system design, performance monitoring, testing, implementation, and maintenance.

13.2 Future Work

There are some issues that have not been fully addressed in this book and may give some directions for future research.

As indicated in Chap. 4, there are two questions to be answered in user interface design for Internet-based control systems. These are how information should be displayed to the remote operators and which information should be displayed. Chapter 4 gives an answer to the first question. The basic problem when designing the user interface for a remote operator is to decide which information to display. One of the main challenges is to design the interface so that the operator is able to quickly realize if certain situations are occurring in the system. A large number of components along with intricate interconnections can congest an interface and visually obscure information that is important at a given point in time. In addition to this, requests for large amounts of information in the interface increases the transmission load over the Internet and limits the scope for concurrency between interfaces requiring a consistent view of common data. As a result, the speed of communication may be slowed down compounding the slower response of the operator if information is obscured (Hussak and Yang 2007).

Real-time data transfer over the Internet described in Chap. 5 is a Java-based solution and employs an end-to-end communication channel and therefore has the same limitations as the Java language. Another promising mechanism for real-time data transfer over the Internet is called Distributed Data Service (DDS), which has been briefly described in Chap. 12. DDS follows a publish/subscribe data model and is based on a data-centric framework. Deadlines, latency, and other timing aspects have been considered in DDS. The obvious advantage of using DDS for real-time data transfer over the Internet is its capability of supporting collaboration among distributed users, control systems, and plants (Wang et al. 2009).

Chapters 6, 7, 8, and 9 discuss overcoming or compensating Internet-transmission latency from both computer network and control theory perspectives. There is a rich literature in this area. Internet-transmission latency is uncertain and unpredictive. Most of the existing research is based on certain assumptions such as the Internet-transmission latency is upper bounded or less than one sampling interval. These assumptions significantly simplify the problem and are an oversimplification of the true Internet scenario. Many control theory researchers are focusing on this topic. More and more publications are being seen in this area (Yang 2006; Huang et al. 2009; Zhao et al. 2009). Given the potential development of the next-generation Internet IPv6 and other enhancements to the WWW infrastructure, the speed of the next-generation Internet might be sufficiently fast to be able to dramatically reduce the transmission delay and data loss. Industrialists will be happy to see that Internet-transmission delay and data loss might become less important in the design of future Internet-based control systems.

SOA, DDS, and Jini Technology have been described as options for implementing the general integrated environment for remote design, testing, and installation for real-time control systems in Chap. 12, but only the Jini technology was used in the implementation. With the rapid development of real-time middleware such as DDS from Real-Time Innovative (RTI) (www.rti.com), using SOA and DDS to implement the integrated environment might become much more easy and efficient. We have used DDS to implement a framework for the integration of geographically distributed hardware and embedded real-time software in one of our projects. The framework can be extended into the remote design, testing, and installation of real-time control systems.

This book focuses on using the public Internet, rather than local networks, such as *fieldbus* and *Ethernet*, as the communication media to carry information flows among sensor, actuator, and controllers. At present, there is a major interest in incorporating wireless networks in real-time control systems. The features of wireless networks are easier installation as no cabling is involved, flexibility, mobility, and efficiency. The most challenging aspects are reliability, interference, and security. Remote monitoring and control over wireless networks are attracting more and more attention in both the control engineering and the computer network communities (Benedetto et al. 2010; Yang and Cao 2008). It is well worth investigating.

References

Benedetto, M.D., Johansson, K.H., Johnson M., and Santucci, F., (2010) Industrial control over wireless networks, *International Journal of Robust and Nonlinear Control*, 20(2), pp. 119–122.

Huang, J., Wang, Y., Yang, S.H., and Xu, Q., (2009) Robust stability conditions for remote SISO DMC controller in networked, *Journal of Process Control*, 19, pp. 743–750.

Hussak, W., and Yang, S.H., (2007) Formal reduction of interfaces to large-scale process control systems, *International Journal of Automation and Computing*, 4(4), pp. 413–421.

Real Time Innovative, available at *www.rti.com*

Wang, Y.H., Yang, S.H., Grigg, A., and Johnson, J., (2009) A DDS based framework for remote integration over the Internet, *Proceedings of the 7th Annual Conference on Systems Engineering Research*, Loughborough.

Yang, T.C., (2006) Networked control system: a brief survey, *IEE Proceedings: Control Theory and Applications*, 7, pp. 537–545.

Yang, S.H., and Cao, Y., (2008) Networked control systems and wireless sensor networks: theories and applications, *International Journal of Systems Science*, 39(11), pp. 1041–1044.

Zhao, Y.B., Liu, G.P., and Rees, D. (2009) Design of a packet-based control framework for networked control systems, *IEEE Transactions on Control Systems Technology*, 17(4), pp. 859–865.

Index

A
Advanced Encryption Standard (AES), 136–139, 142
Advanced process control (APC), 147, 169
Asymptotically stable, 120, 121

C
Collaboration requirements, 185
Compensation of the transmission delay, 121, 123–125
Conceptual design, 170
Control command transmission security, 136–139
Controller performance compensation, 155

D
Damping ratio, 102, 104
Data-centric frameworks, 173, 175–176, 196
Data Distribution Service/Distributed data services (DDS), 171, 176, 178, 192, 196, 197
Data priority, 44, 47, 50, 51
Distributed hardware, 197
Distributed working environments, 172

E
Embedded real-time software, 197
eXtensible Mark-up Language (XML), 37, 39–44, 46–49, 51, 178, 195

F
Fat client, 180–182, 189
Feedback channel, 18, 70, 74, 76–77, 81, 85, 92, 114, 123–124
Feed-forward channel, 18, 73, 74, 77, 78, 80, 81, 85, 92, 124–125, 127

Functional model, 7, 9–12, 15, 139, 171–173, 179

G
General likelihood ratio (GLR), 153–155, 158, 161, 162, 164, 165

H
HAZOP, 140, 142–144
Hierarchical data format (HDF), 37, 39–43, 46–49, 51, 178, 195
Hierarchy of process control, 8, 22, 23, 29, 71–72

I
Information architecture, 12–15
Information hierarchy, 7, 12–13
Intelligent autonomy control, 19, 20, 72, 195
Internet-based control system architecture, 17–26, 69, 195
Internet-based control systems (ICS), 2–3, 7–16, 22, 24–26, 35, 37, 38, 40, 51, 53, 54, 56, 59–62, 64, 67, 68, 72, 74, 85, 92, 95, 99–110, 113–128, 131, 132, 134, 136, 139–142, 144, 147, 187, 195–197
Internet Control Message Protocol (ICMP), 55
Internet load minimization, 101
Internet-transmission latency, 197

J
Java document object model (JDOM), 39–42, 46, 48–51, 195
Java message service (JMS), 176, 178
Jini technology, 176, 178, 179, 197

L

Life cycle model, 170, 171
Light data, 39, 47–49, 51, 157, 165
Linear quadratic regulator (LQR), 116,117, 121, 127
LQG controller, 152–154, 158

M

Malicious hacker, 13, 15, 131, 132, 136
Minimum variance controller (MVC), 151–153
Multi-rate control scheme, 72, 74, 78–79, 95, 96, 100–101
Multi-rate control system, 99, 105, 110, 127
Multi-rate Internet-based control, 99, 110, 113, 121, 128, 196

N

Network-based control, 22
Networked control systems (NCS), 1–4, 17, 18, 72, 99, 110, 128, 195
Network Time Protocol, 56, 195

O

Output feedback, 67, 115–116, 118, 121, 127, 128

P

Performance monitoring, 147–166, 196
Private key, 136, 137
Process control event diagram (PCED), 10–12, 15, 139–142, 144, 196
Process control system hierarchy, 8, 22, 23
Process control unit (PCU), 107, 108, 160, 161, 165
Public key, 136, 137
Public transmission media, 54
Publish/subscribe data model, 196

R

Random round trip transmission time delay, 122
Real-time Control Protocol, 56–59, 195
Real-time data transfer, 37–51, 67, 178, 195, 196
Remote control over the Internet, 4, 21–23, 67, 91, 92

Remote maintenance, 147, 156–157, 161, 162, 166
Remote method invocation (RMI), 42–46, 48, 50–51, 160
Remote monitoring, 166, 197
Requirement specification, 7–16, 26, 67, 195
Rivest–Shamir–Adleman (RSA) algorithm, 136, 137, 139

S

Safeguard, 133–135, 142, 144
Safety interlock system (SIS), 133, 134, 136, 144
Safety risk analysis, 132
Security checking, 131–144, 196
Security risk analysis, 132, 144
Service oriented architecture (SOA), 173–175, 178, 192, 197
Settling time, 99–101, 103, 105, 110, 113
Stability analysis, 104–105, 118–120
State feedback, 117, 118, 121, 123, 128
State feedback control, 114–115, 117, 118, 121–123, 125–127

T

Tele-operation, 3, 17–20, 24, 131, 195
Thin client, 180–182
Time delay compensation, 54, 55, 67, 75, 79–85, 91–96, 121–123, 126
Transmission Control Protocol (TCP), 13, 14, 33, 50, 51, 54–57, 64, 65, 87, 88, 90, 92, 93, 186, 187, 195
Transmission delay, 3, 24, 53–65, 67–96, 100, 113, 114, 116, 121, 123–128, 142, 173, 195–197

U

Unified modelling language (UML), 182
User datagram protocol (UDP), 55–57, 64, 65, 185, 195
User interface design, 29–35, 196

V

Variable sampling time, 69–70, 79–85, 196
Virtual and real controllers, 134, 156, 164, 171–173, 179, 180, 190–192
Virtual and real worlds, 37, 81, 170–173, 178, 185, 188, 192
Virtual machine, 44, 48, 181

Virtual supervision parameter control (VSPC), 68, 69, 85–91, 95, 96, 187, 195
Virtual system, 135, 142, 144

W
Wireless networks, 3, 197
Workbench, 157, 159, 178–183, 185–190, 192

X
XML, 37, 39–44, 46–49, 51, 178, 195

Z
Zero-order-holds (ZOHs), 100, 101

Other titles published in this series (continued):

Soft Sensors for Monitoring and Control of Industrial Processes
Luigi Fortuna, Salvatore Graziani, Alessandro Rizzo and Maria G. Xibilia

Adaptive Voltage Control in Power Systems
Giuseppe Fusco and Mario Russo

Advanced Control of Industrial Processes
Piotr Tatjewski

Process Control Performance Assessment
Andrzej W. Ordys, Damien Uduehi and Michael A. Johnson (Eds.)

Modelling and Analysis of Hybrid Supervisory Systems
Emilia Villani, Paulo E. Miyagi and Robert Valette

Process Control
Jie Bao and Peter L. Lee

Distributed Embedded Control Systems
Matjaž Colnarič, Domen Verber and Wolfgang A. Halang

Precision Motion Control (2nd Ed.)
Tan Kok Kiong, Lee Tong Heng and Huang Sunan

Optimal Control of Wind Energy Systems
Iulian Munteanu, Antoneta Iuliana Bratcu, Nicolaos-Antonio Cutululis and EmilCeangă

Identification of Continuous-time Models from Sampled Data
Hugues Garnier and Liuping Wang (Eds.)

Model-based Process Supervision
Arun K. Samantaray and Belkacem Bouamama

Diagnosis of Process Nonlinearities and Valve Stiction
M.A.A. Shoukat Choudhury, Sirish L. Shah, and Nina F. Thornhill

Magnetic Control of Tokamak Plasmas
Marco Ariola and Alfredo Pironti

Real-time Iterative Learning Control
Jian-Xin Xu, Sanjib K. Panda and Tong H. Lee

Deadlock Resolution in Automated Manufacturing Systems
ZhiWu Li and MengChu Zhou

Model Predictive Control Design and Implementation Using MATLAB®
Liuping Wang

Predictive Functional Control
Jacques Richalet and Donal O'Donovan

Fault-tolerant Flight Control and Guidance Systems
Guillaume Ducard

Fault-tolerant Control Systems
Hassan Noura, Didier Theilliol, Jean-Christophe Ponsart and Abbas Chamseddine

Detection and Diagnosis of Stiction in Control Loops
Mohieddine Jelali and Biao Huang (Eds.)

Stochastic Distribution Control System Design
Lei Guo and Hong Wang

Dry Clutch Control for Automotive Applications
Pietro J. Dolcini, Carlos Canudas-de-Wit and Hubert Béchart

Advanced Control and Supervision of Mineral Processing Plants
Daniel Sbárbaro and René del Villar (Eds.)

Active Braking Control Design for Road Vehicles
Sergio M. Savaresi and Mara Tanelli

Active Control of Flexible Structures
Alberto Cavallo, Giuseppe de Maria, Ciro Natale, and Salvatore Pirozzi

Induction Motor Control Design
Riccardo Marino, Patrizio Tomei and Cristiano M. Verrelli

Fractional-order Systems and Controls
Concepcion A. Monje, YangQuan Chen,
Blas M. Vinagre, Dingyu Xue
and Vincente Feliu

Model Predictive Control of Wastewater Systems
Carlos Ocampo-Martinez

Tandem Cold Metal Rolling Mill Control
John Pittner and Marwan A. Simaan

Control and Monitoring of Chemical Batch Reactors
Fabrizio Caccavale, Mario Iamarino,
Francesco Pierri, and Vincenzo Tufano